Van Tassell

About Island Press

Since 1984, the nonprofit Island Press has been stimulating, shaping, and communicating the ideas that are essential for solving environmental problems worldwide. With more than 800 titles in print and some 40 new releases each year, we are the nation's leading publisher on environmental issues. We identify innovative thinkers and emerging trends in the environmental field. We work with world-renowned experts and authors to develop cross-disciplinary solutions to environmental challenges.

Island Press designs and implements coordinated book publication campaigns in order to communicate our critical messages in print, in person, and online using the latest technologies, programs, and the media. Our goal: to reach targeted audiences—scientists, policymakers, environmental advocates, the media, and concerned citizens—who can and will take action to protect the plants and animals that enrich our world, the ecosystems we need to survive, the water we drink, and the air we breathe.

Island Press gratefully acknowledges the support of its work by the Agua Fund, Inc., The Margaret A. Cargill Foundation, Betsy and Jesse Fink Foundation, The William and Flora Hewlett Foundation, The Kresge Foundation, The Forrest and Frances Lattner Foundation, The Andrew W. Mellon Foundation, The Curtis and Edith Munson Foundation, The Overbrook Foundation, The David and Lucile Packard Foundation, The Summit Foundation, Trust for Architectural Easements, The Winslow Foundation, and other generous donors.

The opinions expressed in this book are those of the author(s) and do not necessarily reflect the views of our donors.

5 Easy Pieces

5 Easy Pieces

How Fishing Impacts Marine Ecosystems

Daniel Pauly

Sea Around Us Project
Fisheries Centre,
University of British Columbia
Vancouver, Canada

THE STATE OF THE WORLD'S OCEAN SERIES

Washington | Covelo | London

5 Easy Pieces: How Fishing Impacts Marine
Ecosystems

© 2010 Daniel Pauly

ISLAND PRESS is a trademark of the Center for
Resource Economics.

Library of Congress Cataloging-in-Publication Data
Pauly, D. (Daniel)
   5 easy pieces : how fishing impacts marine
ecosystems / Daniel Pauly.
      p.   cm.
   Includes bibliographical references and index.
   ISBN-13: 978-1-59726-718-2 (cloth : alk. paper)
   ISBN-10: 1-59726-718-X (cloth : alk. paper)
   ISBN-13: 978-1-59726-719-9 (pbk. : alk. paper)
   ISBN-10: 1-59726-719-8 (pbk. : alk. paper)
   1. Fisheries—Environmental aspects. 2. Marine
ecology. I. Title.
QH545.F53P38 2010
577.7'27—dc22   2009051082

Printed on recycled, acid-free paper ✹

Manufactured in the United States of America
10 9 8 7 6 5 4 3 2 1

Table of Contents

List of Exhibits

Preface

This book features five contributions, originally published in *Nature* and *Science*, in which a demonstration was made of the massive impacts that our modern industrial fisheries have on marine ecosystems. Initially published over an eight-year period, from 1995 to 2003, these five contributions span the transition from the first, initially contested realization that the crisis of fisheries and their underlying ocean ecosystems was global in nature, to its broad acceptance by mainstream scientific and public opinion. The contributions presented in this book contributed to this "mainstreaming," together with numerous others, notably those of J.J. Jackson and his colleagues and of the late R.A. Myers and his colleagues.

Each chapter presents the full text of one contribution, as well as its origin and context, mostly in the form of comments by scientific colleagues, both positive and negative. Also, responses are provided to these comments, and the reception of each of these five contributions in popular media is documented.[1] This provides an opportunity to present, among other things, my views on the extent to which scientists are justified in shaping, by interacting with popular media, the reception of their own, perhaps controversial, results.

Finally, an Epilogue is provided, where some reflections are offered on the manner in which scientific consensus, or rather acceptance, emerges following the presentation of new results, and their discussion within and outside the scientific community.[2]

In addition to the interested lay persons, with whom every author of science-related books would like to engage, I envisage three potential audiences for this small volume. The first is environmental activists interested in the current state of ocean ecosystems and the scientific basis for advocacy aiming to drastically reduce both fishing effort and the influence the fishing industry has on the government agencies that are supposed to control it. The second is graduate students in marine biology and/or fisheries science, who will find in this book a concentrate of issues presently at the core of serious discussions in their disciplines. The third is undergraduates (and their instructors), and perhaps even upper-level high school students (and their teachers), covering the way scientific issues are articulated and debated, as is done in general science courses or in classes devoted to epistemology, that is, critical thinking or theory of knowledge.

With reference to this third group of potential readers, I stress that while attempting to represent the standpoint of critics and other commentators as coherently as possible (with quotes from their work in double column with drop shadow), such as to allow a fair evaluation of their argument, I do not believe that I have achieved impartiality: such blissful state can be reached only by those who do not have anything to say.

The title of this book was undeniably inspired by that of the 1970 film *Five Easy Pieces*, directed by Bob Rafelson and starring Jack Nicholson, and thus, in the last of the endnotes (where one hides peripheral issues, or whimsy prose), I let a professional deal with the parallels between the film and this book.

The five "contributions" (so called to differentiate them from other "papers") introduced, presented in single columns bordered with drop shadows (a style also used for all items from *Nature* or *Science*) and commented upon in this book, are "easy" in the sense that they are based on straightforward analyses of widely available data, and in the sense that the points they make can be understood by anyone—though Chapter 1, which sets the stage, is a bit heavy going and contains details that one might want to skip at first pass. They required, however, a fair bit of data crunching and, occasionally, a new way of looking at the world. Fortunately, this can evolve gradually, step by step, and is thus within the reach of anyone who can read—which is perhaps the main message of this book.

CHAPTER ONE

Primary Production Required

A Summer in Manila

Some people complain about spending too much time commuting. In 1994, during the "International Year of the Family," I started on a transpacific commute that, for five uneasy years, allowed me both to continue living with my family and furthering my research in Manila, Philippines, and to work in my new position as a faculty member of the University of British Columbia, in Vancouver, Canada.[1] In Manila, my employer was the International Center for Living Aquatic Resources Management (ICLARM),[2] which I had joined, fresh from university, in 1979.

As a scientist, I had always straddled two worlds. Although I was born and raised in Europe and was trained at a German university, I had always intended to work in the tropics.[3] This was in no small part because so few of my colleagues had really looked at the tropical fisheries on which so many of the world's poor subsisted. And when they did, they used models and techniques developed for temperate fisheries and fish populations, which had very different characteristics from those of the tropics.[4] But in addition to the geographical worlds I inhabited, I also found myself torn between two scientific worlds: that of theory and that of the application of science to problems in the real world. These tensions proved to be creative ones, and they lie behind the contributions in this book.

ICLARM, in which much of my approach was developed, was, when I joined it, a young organization (it was founded in 1977). It was one of the few research centers devoted to problems of developing-country fisheries, and its scope extended throughout the intertropical belt. ICLARM was a delightful place to work, and the productivity and creativity of its international and national staff were widely recognized (Dizon and Sadorra 1995). Consequently, we were reorganized. By the early 1990s, the bureaucracy had become intolerable, eventually triggering my departure and that of many other colleagues. But not before it had forced upon us at least one positive, if unintended, consequence.

1

One of the purest manifestations of the bureaucratic mummification then taking place at ICLARM was the development of a "strategic plan." In 1990/1991, the entire scientific staff of ICLARM (plus the inevitable consultants) was engaged in developing the plan, which was then supposed to provide guidelines for a midterm plan, which then should provide a framework for annual plans, etc. As a result, we had long discussions on how to evaluate the potential of fish farming (aquaculture) and capture fisheries. Strangely and ironically enough, these discussions inspired the first contribution to this book.

At the time, there was a lot of optimism about the potential of aquaculture, a situation that has not changed, over two decades later. However, the information then available at the eco-regional and global levels did not allow extending this optimism to capture fisheries, notwithstanding their more important contribution to the food security and livelihood of people, notably in developing countries. This did not trouble most fisheries scientists at the time, because for the most part, they didn't look at such data.

This was very much in contrast with agriculture. We noted that the agronomists at the International Rice Research Institute (IRRI) in Los Baños, near Manila, did not write only about their own research plots, or rice culture in the Philippines, their host country, or even that of Southeast Asia, but rather they knew and wrote about global rice production. My colleague Villy Christensen and I found, in comparison, the parochial view of fisheries science odd, since fish were a globally traded commodity and the market saw to it that demand at one place was met with supply from others. We decided to begin to rectify this shortcoming by reviewing the state of, and potential for, catches of fisheries in the entire world. Global fisheries data had been available since 1950, when the Food and Agriculture Organization of the United Nations (FAO) began issuing its admirable global statistical compendia (Figure 1.1).[5]

But only two groups had attempted to produce a global synthesis of the data available at the time: (a) staff of and consultants to FAO, who produced a comprehensive, but already then dated, review composed of chapters by the leading marine biologists and fisheries scientists of the time (Gulland 1970, 1971), and (b) a group led by Moiseev (1969), in the Soviet Union (remember?), whose main conclusions were very different from those then current in the West.[6]

We never formally published our own effort, which was buried as an appendix by Christensen et al. (1991) to ICLARM's very forgettable *Strategic Plan for International Fisheries Research*. It was better so: this review later turned out to have been overly optimistic with regard to the future prospects of fisheries. However, the exercise itself was useful. In particular, it helped us to understand that by examining fisheries at a systemic level, in such a way that we could describe the dynamics of the ecosystem, rather than simply the behavior of any of its component species, we could gain critical new insights.

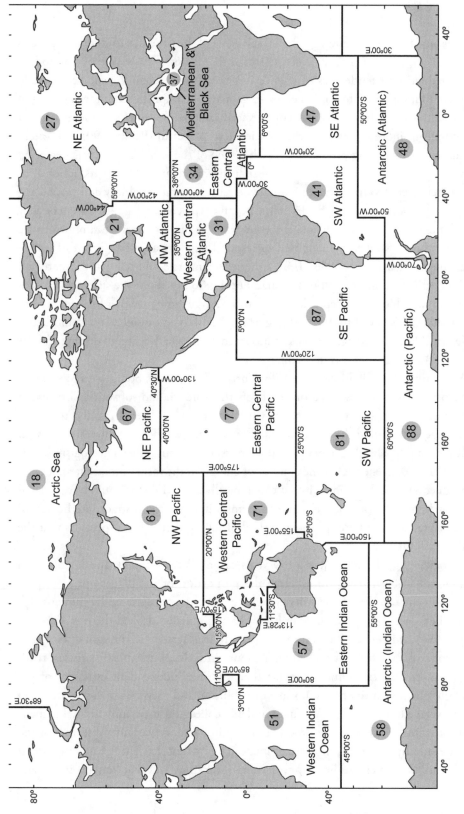

Figure 1.1. FAO Areas, as used by the Food and Agriculture Organization of the United Nations, headquartered in Rome, Italy, to disaggregate catch (i.e., usually landing) data submitted by member countries. These FAO Areas have been little modified since FAO's reporting started in 1950.

The standard practice among fisheries scientists of the time was to study one species and/or one fishery at a time, in isolation from other factors. This was in part a product of the reductionist tendencies of science in general—to isolate the subject of study from confounding variables in an effort to gain an uncompromised understanding of its properties and behavior. This of course should not be viewed as a condemnation of reductionism; indeed, it is what makes science so powerful (Pauly 1990). But often, reductionism causes big problems e.g., when one of the variables that was neglected turns around and bites us. Fisheries research in particular had to be reductionist at first—there were too many factors (notably environmental variability and trophic interactions) to sort out when the discipline began over a century ago. But the tendency of fisheries science to focus on single species was also due to its role in supporting the fishing industry, which was supposed to respond (e.g., by adjusting its effort) to assessments of their target stocks. Thus, fisheries scientists could tell you the status of, say, cod off Newfoundland,[7] but they had nothing to say about the status of the ecosystem on which this cod depended—and this was the same for other fisheries around the world. We believed that by looking at the health of ecosystems, we could better understand how to manage them at a time when catches had begun to decline and the future of many fisheries was in doubt.

At the time, I was working with Villy Christensen on developing ways to summarize ecosystem properties based on the Ecopath approach and software (Christensen and Pauly 1992a), about which there will be more to say later. In trying to model the interactions among the components of ecosystems, we became interested in what H.T. Odum (1988) called "emergy"—not a misspelling, but rather a neologism standing for "embodied energy," or the amount of energy, in the form of food eaten, it takes to produce a particular animal, in our case typically a fish. Emergy could be calculated based on knowledge of food webs in ecosystems, but it was too abstract and theoretical to be of practical use (also, people got tired of telling their word processors that the spelling was OK).

Coincidentally, however, just a few years earlier, Peter Vitousek and his Stanford colleagues had been trying to estimate the proportion of the planet's primary productivity—the capture of the sun's energy by plants—appropriated by humans (for food, fiber, or fuel, or paved over to build shopping malls). I liked the way Vitousek et al. (1986) had derived their estimates for terrestrial systems as the sum of estimates by sector and industry. They relied on a counterintuitive, but robust, statistical truth known as the "central limit theorem": that multiple independent estimates of the same unknown quantity have a normal distribution, and they yield an accurate mean when averaged. Thus, when Vitousek and his colleagues used numerous, independent estimates of the primary production requirements of various subsectors of the global economy, the errors in those estimates largely canceled out upon being added up.[8] The result, which has held up to scrutiny over time, was that humans in the late

---

Box 1.1

Definitions of *Fish, Landings* and *Trophic Level*

Fish: In fisheries science, this term refers to the aquatic animals taken by fishing gears and includes bony fishes (herring, cod, tuna, etc.), cartilaginous fishes (sharks, rays), animals that look like fish but are not (hagfish), and invertebrates (shrimps, crabs, lobsters, oysters, clams, squids, octopi, sea urchins, sea cucumbers, jellyfish, etc.). The term usually excludes marine mammals, reptiles, corals, sponges, and algae, though these are included in some statistics, such as the global fisheries statistics assembled by FAO from annual submissions by its member countries and which provided the starting point for most analyses in this book (Figure 1.1).

Landings: Fishing gears are meant to capture fish by disabling or killing them (von Brandt 1984). Technically, the term *yield* is used for the weight of all fish that are killed (I do not consider, in this book, the fisheries servicing the aquarium trade or other nonfood fisheries), while *catch*, strictly speaking, refers to their number (Holt et al. 1959). In this book, we shall ignore the difference between weight and numbers in catch. Not all fish that are caught are landed and marketed, however. Some are thrown back overboard, and these are called "discards." Thus, one can define Catch = Landings + Discards. Making the distinction between catch and landings is not being pedantic: in the early 1990s, the amount of fish that was discarded annually was estimated at 20–30 million tonnes (t) per year, that is, nearly a third of officially reported landings (Alverson et al. 1994), while more recent estimates put this figure at 7.3 million t (Kelleher 2004; Zeller and Pauly 2005).

Trophic level (TL): This, as I shall elaborate later in this book, is the number of steps in a food web that separates an organism from the primary producers (TL = 1) at the base of that food web.

1980s were appropriating, that is, requiring and/or consuming, 35–40% of terrestrial primary production. By contrast, they found that humans appropriated only 2.2% of marine primary production.

This was an intriguing finding for us. First, primary productivity required (PPR) was just the flip side of Odum's embodied energy concept, and I realized that the "primary production required" by fisheries would itself be a useful, and easily understandable, measure of the impact of humans on marine systems (Christensen and Pauly 1993a; Ulanowicz 1995; Dulvy et al. 2009).

Second, it seemed to me as a fisheries ecologist that Vitousek et al. had not dealt adequately with primary productivity in the oceans. Their estimate of the marine primary production required by fisheries was based on one single multiplication, involving a

---

Box 1.2

Why We Can't Use Mean Trophic Levels to Calculate Primary Production Required

Due to the nonlinearity of the relationships between trophic levels and trophic fluxes, use of mean trophic level (as in Iverson 1990 or Vitousek et al. 1986; see text) for estimating fluxes in multispecies fisheries leads to a strong bias, which can be illustrated using the data in Table 1.1. Inspired by a similar table for the use of terrestrial primary production in Vitousek et al. (1986), it contains estimates of the primary production required (PPR) to sustain the global catches (Y_i) of 39 different groups of fish (i), each calculated using the equation $PPR_i = Y_i (1-TE)^{(TL_i-1)}$ with a mean value for the transfer efficiency (TE) equal to 0.10 (see Figure 1.3), and trophic levels (TL_i) derived from 48 trophic models such as the one in Figure 1.4.

The sum of these estimates is 2.84×10^9 t·year^{-1}. The mean trophic level of these 39 groups, weighted by their catch, is 3.10. This, applied to the sum of the catches, leads, via the above equation, to a PPR estimate of 1.33×10^9 t year^{-1}, or only 47% of the previous estimate that is obtained with disaggregated data. Thus, using mean trophic levels, as in most publications previously reviewed above, completely distorts, through a phenomenon known as "aliasing," the relationships between primary production and catches (or potential yields).

---

then-current estimate of marine fish landings (FAO 1984) times the primary production required to support an "average fish," that is, a fish that would have to be at the exact center of marine food webs (this was defined as having a trophic level of 3.0; see Box 1.1 for definitions of *fish*, *landings*, and *trophic level*).

In reality, this was more complicated. Given the definitions in Box 1.1, one can infer the primary production required to generate a given catch of, say, pollock, if one knows its trophic level and the transfer efficiency of biomass between trophic levels.[9] However, such efficiencies, in marine ecosystems, are rather low (2–20%, with a mean of about 10%; see below). This means that small errors in trophic level assignment will induce large errors in the estimation of primary production required. Thus, the assumption of an "average fish" of trophic level 3.0 in Vitousek's work was misleading because there is no such thing as an "average fish," and even less, one with a mean trophic level representing all fishes (see Box 1.2). Moreover, the trophic levels of most fish were unknown at the time. As it turned out, the solution to this quandary was to be found in Ecopath, a software and modeling approach developed, then abandoned, by Polovina (1984a, 1984b, 1993). It allowed for the first rigorous estimation of trophic levels in aquatic ecosystems, and it had just been rescued from oblivion (Christensen and Pauly 1993a).

Villy Christensen and I used it to reassess the findings of Vitousek et al., and I presented our work at the 1994 meeting of the British Ecological Society in Manchester, where it was viewed by Lawton (1994) as "the most important and disturbing piece of information from the whole conference." I was encouraged to submit a manuscript to *Nature* by Sir (now Lord) Robert May, then already one of the world's leading theoretical ecologists. Actually, at that time, that submission had already occurred, and *Nature*, after providing three easy-to-accommodate referees' reports, not only published our contribution (and listed it on its cover of March 16, 1995), but asked John Beddington, now science advisor to the UK government, to introduce it to its readers (Beddington 1995). This piece is reproduced below, modified only as to its reference style:[10]

NATURE

The Primary Requirements

J.R. Beddington | Vol. 374 No. 6519

Concern that fish are being harvested unsustainably has usually centered on the dramatic collapses of individual stocks or on the conflicts associated with the over-capacity of the fishing industry. At a global level, marked declines in the rate of increase of the total world catch, and indeed falls in that total, have recently prompted the question of whether catches are close to or exceeding the maximum that is sustainable.

Pauly and Christensen (1995) address the issue by calculating the level of primary production necessary to sustain world fish catches. They estimate that some 8% of global aquatic primary productivity is required, a figure that is much higher than previous estimates (Vitousek et al. 1986) but, on the face of it, not particularly large. However, when the contribution of different ecosystems is examined, major differences emerge. For the open ocean, the primary productivity requirement is only around 2%, but open-ocean fisheries contribute only a small proportion of the world fish catch. For the areas of greatest importance to fish production, for instance the continental shelves, the requirements range from 24 to 35%. Such figures imply that current levels of fishing—and certainly any increases—are likely to result in substantial changes in the ecosystems involved.

Pauly and Christensen's calculations start from data collected by the UN Food and Agriculture Organization (FAO) and involve classifying catches of different species into different groups and calculating a fractional trophic level for each group: this statistic is essentially a weighted average of the levels at which species feed. For example, Antarctic krill are given a fractional trophic level of 2.2 on the assumption that they feed 80% at the phytoplankton level and 20% at the herbivorous zooplankton level.

The energy transfer efficiency between trophic levels in aquatic ecosystems is commonly assumed to be 10%. Pauly and Christensen document estimates of transfer efficiency derived from a large number of models which show a mean of around that figure, which they therefore adopt. The energy requirement to sustain fish catches can then simply be compared to estimates of primary productivity for different ecosystems. The figures calculated are subject to a number of queries; as the authors point out, estimates of primary productivity are on the low side, but by contrast catches by artisanal and subsistence fishermen are probably under-reported. Nonetheless, such considerations are unlikely to alter the conclusions of the [contribution] a great deal.

A notable complication that Pauly and Christensen take into account is the amount of fish caught and discarded for economic reasons. The annual level of discards has been estimated to be some 27 million tonnes (Alverson et al. 1994), a staggering figure in the context of a total world catch of less than 100 million tonnes.

One potential snag is that the analyses critically depend on the catch composition of the fisheries, and in particular on their trophic level. The authors avoided this problem by calculating average catches from the period 1989–91, when the species composition of the world catch did not alter markedly. However, there is a more important dimension to this question. If the level of fishing means that too high a proportion of primary productivity is being required, then it is likely that the ecosystems will move towards a composition increasingly dominated by lower trophic levels.[11] This will tend to lower the primary production requirements.

It is well known that sustainable yield from aquatic ecosystems can be increased by harvesting species of lower trophic levels—there is the obvious gain to be made from saving in energy transfer; and species at lower trophic levels typically have somewhat higher ratios of productivity to biomass, and hence higher sustainable yields. Such a change is a probable consequence of fishing practice in any case. The shift in emphasis towards lower trophic levels can be intentional or a natural consequence of overfishing of species at higher trophic levels. The first situation is likely to occur when the economically preferred species is at a lower trophic level (for example, shrimp). The second, and probably more common, situation occurs when the preferred species is from a higher level (for example, cod).

Pauly and Christensen's estimates provide a snapshot of how things stood at the turn of the 1980s. Major changes in catch and species composition in aquatic ecosystems have certainly occurred in response to fishing. They are likely to continue, but whether primary productivity requirements will increase or decrease as a result is unclear. What is clear is that an unlimited movement down the trophic web in target species cannot continue without causing economic difficulties, losses in biodiversity and conservation problems for higher species in the ecosystems, including marine mammals. The importance of the work is to draw attention to these limits and to highlight the necessity of monitoring the relevant statistics at a global level.

The limit to which complexes of interacting species may be exploited is still an open question. Ignorance on that score is less of a drawback than it might appear, as the scientific issues of assessing the sustainable yield of individual fish stocks are relatively

NATURE

well understood.[12] The limit to which a single species can be exploited as a proportion of its biomass is known to be a simple function of the parameters of growth and mortality. When considering harvesting at different levels in the ecosystem, it has been shown that, for a wide range of life-history characteristics, yield is proportional to natural mortality (Kirkwood et al. 1994), which itself tends to be inversely related to trophic level. Sustainable harvesting of fluctuating stocks can be addressed by feedback policies, and in principle it should be possible to manage fisheries sustainably.

Principle is one thing, however, practice another. The main force driving the level of exploitation to the unsustainable is overcapitalization of the fishing industry, which itself has been produced by the open-access nature of fisheries. The FAO has estimated annual losses of the global fleet to be some US$54 billion. These losses are largely met by subsidies. The failure of fisheries management to address the problems of unrestricted access and to stop unsustainable levels of harvesting (which have led to the depletion of many major stocks) is estimated to be costing a further US$25 billion each year. These figures show the potential benefits to be had from successful fisheries management on the world scale.

The contribution in question now follows, modified with regard to its reference style, and also incorporating the few inconsequential corrections later published in an erratum.

NATURE

Primary Production Required to Sustain Global Fisheries[13]

D. Pauly and V. Christensen | Vol. 374

The mean of reported annual world fisheries catches for 1988–1991 (94.3 million t)[14] was split into 39 species groups, to which fractional trophic levels, ranging from 1.0 (edible algae) to 4.2 (tunas), were assigned, based on 48 published trophic models, providing a global coverage of six major aquatic ecosystem types. The primary production required to sustain each group of species was then computed based on mean energy transfer efficiency between trophic levels of 10%, a value that was re-estimated rather than assumed. The primary production required to sustain the reported catches, plus 27 million t of discarded bycatch, amounted to 8.0% of global aquatic primary production, nearly four times the previous estimate. By ecosystem type, the requirements were only 2% for open ocean systems, but ranged from 24 to 35% in fresh water, upwelling and shelf systems, justifying current concerns for sustainability and biodiversity.

Global primary productivity generates annually about 224 x 10^9 t dry weight of biomass. Of this, 59% is produced in terrestrial ecosystems, the rest in aquatic systems

NATURE

(Vitousek et al. 1986). Of the terrestrial primary production, 35–40% is presently used by humans, directly (for example, as food or fiber), indirectly (for example, as feed for animals) or foregone (through, for example, urban sprawl; Vitousek et al. 1986). This was estimated by adding the primary production required (PPR) by various production systems (such as cultivated and grazing lands) or by various sub sectors (such as timber or fibre production), thus allowing errors in the independent subtotals to cancel out partly.

The PPR to sustain the world's catches of 75 million t in the early 1980s was also estimated in the same study, based on the key assumption that the "average fish" feeds two trophic levels above the primary producers. This suggested that 2.2% of the world's aquatic primary production was required to sustain the fisheries, and thus led to the conclusion that "human influence on the lowest trophic levels in the ocean (outside severely polluted areas) is minimal, and human exploitation of marine resources therefore seems insufficient by itself to alter on a large scale any but the target populations and those of other species interacting closely with target species" (Vitousek et al. 1986).

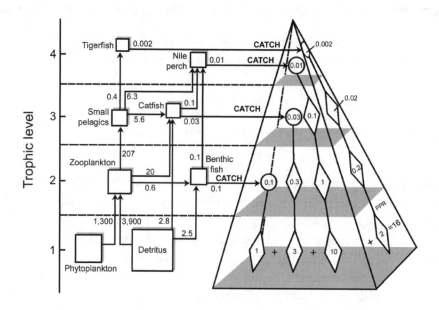

Figure 1.2. Schematic representation of approach used here to estimate the PPR to sustain the catches of a given ecosystem. Left: the simplest among the 48 trophic models used here, representing the lightly fished Lake Turkana (Kolding 1993). Each of its state variables (boxes) has inputs (food) and outputs (predation and/or fishery catches), in t (wet weight) km^{-2} year^{-1}; only major flows are shown, excluding respiration and back flows to the detritus. Right: the pyramid illustrates how the catches (circles) are raised to PPR at trophic level 1 (diamonds), using the 10% transfer efficiency of Figure 1.3.

This work is an attempt to obtain a more accurate estimate of the PPR to sustain the world fisheries catches (including discarded "bycatch"), based on the same approach as used above to estimate terrestrial PPR, wherein independent estimates are obtained on a commodity group and system basis, then added up to yield a robust estimate of the total.

The approach illustrated in Figure 1.2 uses only flows of matter (catches and food consumption of fishes and their prey) and does not require estimation of biomasses, which have proved hard to estimate reliably on a global basis (Gulland 1971).

Recent world fisheries statistics, covering a short period (1988–1991) without major changes in catch composition and reported by the UN Food and Agriculture Organization (FAO 1993b), were split into 39 commodity groups, by ecosystem types. The PPR was then estimated by group, and ecosystem type, based on an estimate of 10% mean transfer efficiency between trophic levels (Figure 1.3), and the mean trophic levels of the commodity groups (Table 1.1). This led to group-specific estimates of PPR, presented here by ecosystem type, after accounting for the 27 million t of discarded

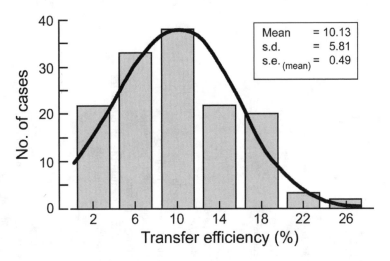

Figure 1.3. Frequency distribution of energy transfer efficiencies (in %) in 48 trophic models of aquatic ecosystems. The estimates of transfer efficiency (N = 140) express, for trophic level 2 (= herbivores and detritivores) to 4 (= third-order consumers), the fraction of production passing from one trophic level to the next, and account for consumers feeding on the different trophic levels of an ecosystem (Cousins 1985; Christensen and Pauly 1992a). No trend of transfer efficiency with trophic level was apparent (Christensen and Pauly 1993a). All 48 trophic models used as sources of transfer efficiencies are fully documented (Christensen and Pauly 1992a; Christensen and Pauly 1993b; Pauly and Christensen 1993; Christensen 1995b; Jarre-Teichmann and Christensen 1998); they jointly show that the 10% transfer efficiency value commonly used for aquatic organisms (May 1976) is extremely close to the mean of the 140 estimates available for this study.

Table 1.1

Reported world fisheries catches, ancillary statistics and the PPR to sustain these catches

FAO-codes	Species group	Catch (ww; t x 10³)	n	k	Trophic level Mean	s.e.	PPR (g C x 10¹²)
Oceanic (gyre) systems							
36	Tunas, bonitos, billfishes	2,975	1	3	4.2	0.04	523.9
46	Krill	344	—	—	2.2*	—	0.6
Upwelling systems							
35	Anchovies, sardines	11,597	24	97	2.6	0.28	53.1
34	Jacks	4,785	8	28	3.2	0.06	86.7
37	Mackerels	1,096	10	44	3.3	0.10	22.8
57	Squids†	248	6	31	3.2	0.14	6.9
Tropical shelves							
24, 35	Small pelagics	7,127	5	20	2.8	0.27	59.9
31, 33, 39	Misc. teleosteans	5,342	22	16	3.5	0.26	204.3
34, 37	Jacks, mackerels	2,053	8	46	3.3	0.28	45.5
36	Tunas, bonitos, billfishes	1,275	8	44	4.0	0.12	141.7
57	Squids, cuttlefishes, octopuses	1,114	6	31	3.2	0.14	19.6
45	Shrimps, prawns	650	4	21	2.7	0.35	35.0
42–44, 47, 77	Lobster, crabs and other invertebrates	544	7	35	2.6	0.30	2.2
38	Sharks, rays, chimaeras	344	9	51	3.6	0.24	15.2
Non-tropical shelves							
32	Cods, hakes, haddocks	12,209	5	49	3.8	0.25	929.9
33	Redfishes, basses, congers	3,837	2	5	3.4	0.06	110.9
39	Miscellaneous marine fishes	3,362	1	5	3.2	0.11	52.8
34	Jacks, mullets, sauries	2,871	1	3	3.8	0.13	206.0
35	Herrings, sardines, anchovies	2,319	3	8	3.0	0.15	23.7
42–45, 47, 75, 77	Shrimps and other crustaceans	1,195	3	10	2.3	0.24	2.6

57	Squids, cuttlefishes, octopuses†	1,114	6	31	3.2	0.14	19.3
31	Flounders, halibuts, soles	1,098	3	10	2.9	0.12	9.8
37	Mackerels, cutlassfishes	1,096	3	16	3.4	0.29	30.6
23–25	Diadromous fishes	819	14	49	2.4	0.25	2.3
38	Sharks, rays, chimaeras	344	2	15	3.7	0.28	19.2
Coastal and coral systems							
52–56, 58	Bivalves and other mollusks	5,150	4	12	2.1	0.13	7.6
31, 39	Miscellaneous marine fishes	3,424	15	86	2.8	0.41	24.0
35	Herrings, sardines, anchovies	2,319	9	52	3.2	0.20	40.8
9	Seaweeds	1,683	1	—	1.0	—	0.2
34, 37	Jacks and mackerels	1,322	17	97	3.3	0.22	29.3
23–25	Diadromous fishes†	819	3	13	2.8	0.19	5.7
43–45, 47	Shrimps, prawns	748	8	42	2.6	0.33	3.3
42, 74–77	Crustaceans and other invertebrates	566	14	49	2.4	0.25	1.6
72	Turtles	2	2	7	2.4	0.37	0.006
Freshwater systems							
13	Misc. freshwater fishes	5,237	41	273	3.1	0.28	69.4
21–25	Misc. diadromous fishes	1,210	23	121	3.6	0.27	60.1
41, 45, 51, 54, 71, 77	Invertebrates and amphibians	896	14	54	2.2	0.23	1.6
11	Carp-like fish	632	15	79	2.7	0.34	3.7
12	Tilapias and other cichlids	579	24	11	2.5	0.18	2.0

The items used to infer primary production required (PPR) from the reported annual catches (wet weight) from the reported annual catches (wet weight; means for 1988–1991) were: (1) codes to link the groups in the FAO catch statistics (FAO 1993a) to those in the 48 trophic models of ecosystems whose sources are given in Figure 1.3; (2) major group names; (3) group catch (total average catches less aquaculture production for 1988 of freshwater, brackish water and marine fish fed on artificial fish farms, except for India, whose annual production of carp-like fishes was assumed at 2×10^5 t), assigned to ecosystem type based on ecological and geographic considerations; (4) mean trophic level (TL), estimated for the groups included in the models in Figure 1.3, with standard errors (s.e., in brackets) estimated from: $\sum_{i=1}^{k} s_i^2 (n_i - 1) / (n - k)$ where n_i is the number of prey items used for one estimate of TL, n the sum of all n_i, k the number of TL estimates used to compute each group's mean TL, and the s_i^2 are the variances of these TL estimates, that is, the indices of omnivory (Christensen and Pauly 1992a); (5) values of n_i and k. The PPR estimates are based on a conservative 9:1 ratio for the conversion of wet weight to carbon (Strathmann 1967) and a 10% transfer efficiency per trophic level (Figure 1.3), that is, using PPR = $(\text{catches}/9) \times 10^{(TL-1)}$.

*Assumed diet composition 80% phytoplankton, 20% herbivorous zooplankton (McClatchie 1988).
†No system specific estimate of trophic level available; instead a value from another system type is used.

bycatch recently estimated, based on thorough review of worldwide discarding practices (Alverson et al. 1994; see Table 1.2).

The results differ markedly from those of the previous study based on an "average fish": we estimate that 8.0% of the world's aquatic primary production is required to sustain the fisheries, nearly four times the earlier estimate. The difference is due to having used higher fisheries catches, considered discards, and relying on disaggregated data, to account for the non-linearity of the relationship between PPR and trophic levels. Although the 8% still may be a moderate figure compared to 35–40% for the terrestrial systems, the prospects for increases are dim. The bulk of aquatic productivity

Table 1.2

Global estimates of primary production (PP), of PPR to sustain world fisheries (mean for 1988–1991, wet weight), and of the mean trophic levels (TL) of the catches, by ecosystem type

Ecosystem type	Area (10^6 km²)	PP (gC·m⁻² year⁻¹)	Catch (g·m⁻² year⁻¹)	Discards (g·m⁻² year⁻¹)	TL of catch	PPR (catches + discards) Mean (%)	PPR (catches + discards) 95% Confidence interval
Open ocean	320.0	103	0.01	0.002	4.0	1.8	1.3–2.7
Upwellings	0.8	973	22.2	3.36	2.8	25.1	17.8–47.9
Tropical shelves	8.6	310	2.2	0.671	3.3	24.2	16.1–48.8
Non-tropical shelves	18.4	310	1.6	0.706	3.5	35.3	19.2–85.5
Coastal/reef systems	2.0	890	8.0	2.51	2.5	8.3	5.4–19.8
Rivers and lakes	2.0	290	4.3	n.a.	3.0	23.6	11.3–62.9
Weighted means (or total)	(368.8)	126	0.26	0.07	2.8	8.0	6.3–14.4

Distribution of surface areas by ecosystem type was estimated based on planimetry, checked against published estimates (Lieth 1978). As defined here, coastal systems generally reach down to a depth of 10 m, except coral reefs which may reach to about 30 m (Crossland et al. 1991). The PP estimates are based on de Vooys (1979), with a 1:3 allocation between ocean and shelf productivity (Strathmann 1967), and using newer, higher values for upwelling systems, leading to a more conservative estimate of PPR. The nutrients released by discarded fish are assumed to have negligible effects on primary production and on food webs in general. The catches are from Table 1.1 and their trophic levels (TL) are weighted means. The PPR estimates are from Table 1.1; they were adjusted to account for the discards (allotted across systems based on group-specific catch/discard ratios; Alverson et al. 1994). Their standard errors (s.e.) were estimated by the Monte Carlo method assuming normal distributions, with the s.e. of the mean transfer efficiencies in Figure 1.3, and of the mean TL in Table 1.1 providing all the variability (10,000 runs per group in Table 1.1, all within ± 1 s.e. of each mean), the FAO catches, the discards, and the PP by ecosystem type being used as fixed values, notwithstanding their imprecision (see text).

(75%) occurs in the open ocean (gyre) systems, by virtue of their vast extent. Only 1.8% of this productivity is used, but as little as 20 to 25% of the overall zooplankton biomass may be available for the higher trophic levels (Roger 1994), dominated by top predators (notably yellowfin and skipjack tuna), which must roam desert-like ocean expanses to find scattered food patches.

The estimated PPR for the coastal and coral reef systems is 8.3%. This relatively low value is due to (1) a high level of productivity (Table 1.2), (2) large catches at low trophic levels (seaweeds, bivalves and other invertebrates), and (3) overfishing, which has left the reduced fish biomass unable to use the available production.

The shelf systems exhibit high PPR, from 24.2 to 35.3%, mainly due to industrialized fisheries operating at high trophic levels, a feature in which they differ from upwelling systems, for which, however, the PPR is still high, 25.1% (Table 1.2). It would seem difficult to further increase these values, especially on temperate shelves, given that a substantial part of the primary production generated during intensive blooms settles out as detritus, before use by zooplankton is possible. Also, higher PPR would starve top predators, such as marine mammals and birds.

Could catches be increased through fishing down the food web, that is concentrating on fishes at lower trophic levels? In ocean systems, the fisheries targeting tuna would have to harvest the prey of tuna (mainly small pelagic fish), which cannot be done economically at present. In coastal and coral reef systems, fishing has already moved down the food web, and improvements must come from rebuilding biomasses through better management (May et al. 1979; Pauly 1986b). This is also true for tropical shelves, where intensive overfishing is causing significant loss of spawning biomass and of biodiversity, especially through shrimp trawling on soft-bottoms, which results in destruction of soft corals and massive changes in community structure, including large fish being replaced by short-lived organisms (small pelagic fish, cephalopods, jellyfish and so on) (Pauly 1986b). This is aggravated by the fact that, contrary to some terrestrial ecosystems such as rainforests, of which large undisturbed tracts still exist, and contrary to what is stated in the introductory quote, the overwhelming bulk of the world's trawlable shelves are impacted by fishing, leaving few sanctuaries where biomasses and biodiversity remain high.

There is at present a debate about the level of global primary production, which may be slightly higher than the figure we have used (Li et al. 1983; Post et al. 1992). However, the fisheries catches are likely to be underestimated as well, despite having been adjusted for discarding practices, because of under-reporting and under-collection, assumed by numerous fisheries scientists, but still awaiting the kind of global analysis now done for discards (Alverson et al. 1994).

The results, having been obtained by summing independent group- and system-specific catches and PPR estimates, should be robust. Moreover, the estimate of PPR of 8% for all aquatic systems, although nearly four times as high as the previous estimate, masks the important fact that the nearshore systems readily accessible to humans (most upwellings and the shelves), and the freshwater systems have very high PPR, nearing that estimated for terrestrial systems. Because we do not have the ability safely to increase aquatic primary production, these high PPR values confirm broad limits on the carrying capacity of natural aquatic ecosystems, which still form the basis for 85% of the world fish harvest.

A Response and a Tedious Rejoinder

I was mightily pleased when the above contribution appeared and basked for a few days in the afterglow of colleagues' congratulations (such as you get when you first publish in either *Nature* or *Science*). However, I was soon brought back to earth by a tersely worded message from the editor of *Nature* instructing me, as first author, to start preparing a response to a letter they had received that questioned our use of an estimate of 10% as an appropriate mean value for transfer efficiency within the world's ocean ecosystems.

Nature, however, has such letters refereed anonymously, as it does original contributions, and apparently the referees of the letter in question thought it not worthwhile: the challenge was not published by *Nature*. Rather, although it had been originally signed by Professor Timothy Parsons (then at the University of British Columbia) and one of his graduate students, it appeared in a journal edited by Prof. Parsons, under the sole authorship of that graduate student (Baumann 1995). I include it here not as a straw man, which in some sense it is, but rather because it reveals a peculiar set of assumptions that at the time were nearly beyond question. As for the draft of the now superfluous response, I incorporated it into the body of a contribution (Pauly 1996; see Appendix 1) that the editor of another journal, *Fisheries Research*, happened to solicit just at that time.

The first debate thus documents the arguments in Baumann (1995), who challenged the choice of a value for the mean transfer efficiency of marine ecosystems, and my response:[15]

A Comment on Transfer Efficiencies
M. Baumann | Vol. 4, No. 3

The gross growth efficiency, defined as growth divided by ration, is very high for aquatic organisms, including both plankton and fish. If we assume that animals in nature have to maintain a rapid growth rate in order to survive, then growth efficiency values in the range of 20–40% are common (Hoar et al. 1979; Parsons et al. 1984). [...]. It is surprising, therefore, that the most popular value for transfer efficiency is

10% as summarized from a survey of the literature by Pauly and Christensen (1995). While one would expect the transfer efficiency between trophic levels to be lower than the growth efficiency, choosing the most popular value of 10% is not a justification of its validity. However, taking a mean growth efficiency of 30% it becomes clear that the predation efficiency [...] is critical in determining [transfer efficiency]. Pauly and Christensen (1995) try to provide a robust estimate of the primary production required (PPR) to sustain world fisheries catches by employing [a] simple equation [which includes...] a constant transfer efficiency between trophic levels of 10% [...]. This was not re-estimated, as claimed by Pauly and Christensen (1995); rather, they took a mean of transfer efficiency model assumptions.[16] A mean calculated from assumptions is still an assumption, unless accuracy is determined by a popular belief.

Furthermore, Pauly and Christensen (1995) apply Monte-Carlo simulations to estimate the 95% confidence intervals of PPR for the different ecosystems using the standard error of the transfer efficiency assumptions to capture the variability of transfer efficiency within different species groups. Nevertheless, from the literature (Sheldon et al. 1977; Iverson 1990; Parsons and Chen 1994) it appears that the transfer efficiency of carbon in the sea tends to be closer to 15%, which makes the simulations meaningless. Changing the energy transfer efficiency from 10% to 15% reduces the primary productivity required to sustain world fish catch from 7.5% to 2.6% in my calculations, which is close to the value of 2.2% reported earlier (Vitousek et al. 1986). Although the Monte-Carlo simulations reflect the variability of trophic levels among species groups they ignore the variability in trophic levels within a group.

Baumann then concluded with the observation that our method

includes several assumptions. First, it is assumed that fish production is limited by factors that control carbon transfer. Iverson (1990) showed that fish production is more likely controlled by the amount of new nitrogen passed through the food chain. Nitrogen tends to be conserved in marine systems and has transfer efficiency of up to 28%. Second, transfer efficiency remains constant along the food chain. This would

be really surprising considering the amount of effort higher trophic levels have to invest to find and catch food. Third, transfer efficiency is independent of length of the food chain. However, a general pattern of decrease in mean transfer efficiency with increase in the number of trophic levels has been established (Ryther 1969). Iverson (1990) even inferred a functional relationship between them.

It's important to understand the context for this critique. The transfer efficiencies suggested by Baumann had been used in a number of estimates of the biomass of high-trophic-level fish (the kind we generally like to eat). To reiterate the method: by using estimates of transfer efficiency and primary production, one could estimate the amount of energy moving from one trophic level to the next, based on assumptions about predation of organisms at each level, which then yields estimates of the biomass at each trophic level. (Of course, this wasn't the only method used; other estimates had been derived by extrapolating catch trend and extrapolation from known areas to the global ocean, but it is the one relevant to the dispute here.) Beginning with Graham and Edwards in 1962 and continuing with Schaefer in 1965 (who in 1954 proposed the famous "Schaefer model," which any fisheries student is supposed to know) and Ryther in 1969, among many others, fisheries scientists, at least in the West, had used this method to estimate that the potential yield of marine fisheries hovered about a figure of 100 million tonnes annually, substantially higher than the then-annual catch of 75 million tonnes.

Typical of the assumptions that went into such estimates was this, from a 1969 paper by Ryther: "Slobodkin (1961) concludes that an ecological efficiency of about 10% is possible and Schaeffer [sic] feels that the figure may be as high as 20%. Here therefore, I assign efficiencies of 10, 15 and 20 percent, respectively, to the oceanic, the coastal and the upwelling provinces, though it is quite possible that the actual values are considerably lower."

In other words, Ryther, favorably cited by Baumann, *guessed* the values of the key factors of his estimate of the potential yield from the ocean, but he obtained a value found acceptable by his contemporaries (Ryther's paper was, and still is, well cited). A step-by-step review of the scientific tradition that made this possible is provided in Appendix 1, documenting that, whatever their starting points, the scientists involved in this business unfailingly concluded that approximately 100 million tonnes of fish were available to us to catch sustainably.

Thus, the interesting point about Baumann's critique of our paper was, not his (incorrect) characterization of the methods by which we estimated transfer efficiency, but rather the fact that his own analysis relied on assumptions that had so often been repeated in the literature that they were taken at the time to be true. There is no need to pick on Baumann here, as one of the great challenges in evaluating the scientific literature is the difficulty of distinguishing dead bones transferred from one grave to the other from sound estimates based on real data. In the case of estimating transfer efficiencies, these analysts almost always assumed that a 10% transfer efficiency was conservative and used higher numbers, leading to higher estimates of the total biomass available at higher trophic levels.

What's even more interesting is one particularly bold (and inaccurate) assumption—that in an optimally exploited ecosystem, half of the natural fish production should be left for predators, and half taken by fisheries. It first appears in the 1962 paper from Graham and Edwards (which not coincidentally was the first to suggest a total fisheries potential of 100 million tonnes): "Properly harvested, it is reasonable to suggest that resources of this nature may yield 50 per cent by weight, at last, of the net annual production."

This "reasonable" suggestion turned out to have far-reaching consequences. In the preface of an FAO review edited by him, and widely circulated in book form (Gulland 1971), John Gulland presented the derivation of the (in)famous "Gulland equation"—whose true originators appear to be Alverson and Pereyra (1969). It has the form "Potential yield = $0.5 \cdot M \cdot B_0$" where M is the natural mortality of the stock in question and B_0 its unexploited biomass. This was based on two sets of arguments:

First, the surplus production model of Schaefer (1954) implies that unexploited biomass is halved when maximum sustainable yield (MSY) is achieved. Thus, the fishing mortality generating MSY (F_{MSY}) is about equal to natural mortality (M), and the logic in Gulland's equation applies.

Second, yield per recruit (Beverton and Holt 1957, 1964) is optimized for values of mean lengths at first capture ranging from 40 to 70% of the maximum length of the fish in question,[17] when fishing mortality is about equal to natural mortality, or $F_{opt} \approx M$.

Given that the ratio of production to overall biomass (P/B) is equivalent to total mortality (Z) and that $Z = F + M$ for standard representations of growth and mortality (Allen 1971), Gulland's equation implies that MSY represents 50% of biological production—precisely the assumption of Graham and Edwards (1962; see above), which Gulland, curiously enough, does not cite.

Many subsequent authors built on this assumption (see, e.g., Dickie 1972; Parsons and Chen 1994), but dedicated studies (Francis 1974; Beddington and Cooke 1983; Kirkwood et al. 1994; Christensen 1996) show it to be untenable: the fishing mortality that maximizes sustainable yield is, for most single- or multispecies fish stocks, much smaller than M, or $F_{MSY} \approx 0.2 - 0.5 \cdot M$. The implications for potential yield estimates that assume $F_{MSY} \approx M$ are obvious (you will overestimate potential yield!). Also, as noted by L. Alverson (personal communication), it is not obvious that M at B_0 is equal to M at $B_0/2$.

Now, I will be the first to admit that this rejoinder is a bit tedious. But it's important to note how, even among respected scientists, assumptions have a way of perpetuating themselves, even if they are based on nothing but guesses. There is no basis for believing 100 million tonnes to be "the" annual world catch (see Figure 4.1 for another reason not to believe it) or its potential catch if fisheries were well managed. Yet the dubious origin of the 100 million tonnes estimate (see Appendix 1) was eventually forgotten in the fisheries world, and this estimate is still accepted as fact in some quarters. It is thus fitting that Bjørn Lomborg, the self-styled "skeptical environmentalist," would fall into this 100 million tonnes hole, then dig himself deeper by declaring that catching only 90 million tonnes per year (as opposed to 100) is a (small) price we pay for overfishing.[18] He really wrote this.[19]

The World According to Pimm

Stuart Pimm is an ecologist par excellence and he once took on the job of auditing the entire Earth, in a book called *The World According to Pimm* (2001), in which he recalculated how much of its primary production is appropriated by humans. Pimm's interest in this topic stems from his longtime association with population biologist Paul Ehrlich of Stanford University, who was a coauthor of the Vitousek et al. (1986) publication that served as the model for our contribution. Pimm describes one of their encounters, in Hawaii, as follows: "Paul Ehrlich and I looked out from the dark green forest of the Alaka'i swamp 1000 m down the pali to the ocean. 'What percentage of the ocean's productivity does humanity use?' was the question on our minds."

The answer to their question was long under way, but part of the foundation was laid at a conference in Berlin in 1990 (see Gordon et al. 1991):

The World According to Pimm

Five years later I am in Berlin for a conference [... and] we were debating an editorial in *The Lancet*, the prestigious British medical journal. It dealt with the consequences of sterile oral rehydration packets. These are sugars and salts that children need if they are to survive dysentery and related diseases. Worldwide, most children do not get the help they need, and those diseases are leading causes of human death. The gist of the editorial was "What happens when these children grow up?" "Do they then not starve to death in even greater numbers? Is allowing the children to die the better choice?"

The doctors felt they had an easy answer to this question. It is a question no one can answer, least of all people professionally committed to saving lives. "Use more of the planet to grow food!" they recommended. [...] The doctors were unaware that we had already co-opted a great deal of the planet's annual plant growth or that little useful land remained. They thought that expanding the area of agricultural land would be a cinch.

"Then the solution to our dilemma is the oceans," one doctor ventured. "Aren't there still plenty of fish in the sea?" I could not answer the question easily, so I looked to the other ecologist in the group, Daniel Pauly, a fisheries ecologist then working in the Philippines.

[The] answer, then in embryonic

form, was elaborated some years later in the pages of *Nature*. Working with Danish marine ecologist Villy Christensen, Pauly produced two calculations for the oceans to parallel those of the Stanford group for the land.

Their first calculation is of how much green stuff the oceans produce. Unlike the land, the oceans' plants are phytoplankton. The method of scissors and undergraduates will not work for them, except along the ocean fringe where seaweed grows in shallow water. The world's oceans are bigger, deeper, and far more difficult to explore than is the land.

How do we know where to take the measurements? The ocean, like the land, has its barren deserts—the blue waters far from land—and its productive oases, the green seas. I needed to talk to ecologists who knew about the two-thirds of the planet that is completely unknown to me. To find them I had to enter their domain: the ocean and the research ships on which they visit it. [...]

Pauly and Christensen's second calculation was how much of the oceans' green stuff we eat. Unlike the land, most of what we take from the oceans is not plant life, nor even animals that eat plants, but the ecological equivalent of lions and tigers: animals that eat the animals that eat the plants. They calculated how much of the oceans' green stuff feeds the small animals that feed the fish that feed us. [...]

Whether on land or in the sea, plants are green. The blue oceans produce very little green stuff and what they do produce is scattered over vast areas; we can use little of it. Green seas are mostly coastal and shallow. Pauly and Christensen's numbers show that a third of the oceans' production goes to support our fisheries—a figure broadly similar to the two-fifths of the land's production. That is a large percentage and certainly does not offer unlimited hope for oceans to be a new source of food for humanity. Nor is it one that suggests that the oceans are unaffected by our activities. [...]

To get the numbers of who eats whom, by how much, and with what efficiency, Pauly and Christensen examined 48 detailed case studies and obtained 140 estimates of efficiencies. Their average was close to 10 percent, and nearly two-thirds of the values were between 6 and 14 percent.

The second part of their work was to calculate an average trophic position for each type of fish. Pauly and Christensen estimated that tuna and some similar fish get 80 percent of their food from the third trophic level, which is fish that eat zooplankton that eat phytoplankton. The remaining 20 percent comes from the fourth trophic level: the fish that eat fish that eat zooplankton that eat phytoplankton.

The final stages of the process were already done by the FAO. Their fishery statistics group all the 1000 or so kinds of fish that we catch into 37 major categories on the basis of their natural histories. One of these categories includes bonito, wahoo, mackerel, tuna, sailfish, marlin, and swordfish. Some of the catches are large (skipjack tuna and yellowfin tuna, for example, account for about 3 million tons of catch per year).[20] Others in this grouping have landings only a thousandth of this size.

The fishery statistics also group these catches into geographic regions, for example, the eastern tropical Pacific, the western tropical Pacific, and so on. [...] The catches of yellowfin tuna, for instance, come almost entirely from the eastern and western tropical Pacific and the tropical Indian Ocean. These are all tropical open ocean systems.

We can now [get our numbers]. Some 0.73 million metric tons (dry weight) of tuna and similar fish (including bonito, wahoo, mackerel, sailfish, marlin, and swordfish) are caught in open ocean systems. Another 0.30 million tons dry weight come from tropical shelves.

Suppose that all these fish fed at exactly trophic level four:

0.73 million tons of bonito, wahoo, mackerel, tuna, sailfish, marlin, and swordfish would eat:

7.3 million tons of smaller fish, which eat:

73 million tons of zooplankton, which eat:

730 million, or 0.73 billion, tons of phytoplankton from the open ocean.

Pauly and Christensen's databases showed that this is too simple a formula, because these fish derive 20 percent of their food from the next higher level. So, the previous numbers are 80 percent correct (which comes to 0.58 billion tons of phytoplankton). To that we must add 20 percent of 7.3 billion tons of phytoplankton. If these fish got all their food at this higher level, then 7.3 billion tons of phytoplankton production would be needed to support the 0.73 billion tons of zooplankton production, the 73 million tons of smaller fish, the 7.3 million tons of predatory fish, and the 0.73 million tons of tuna. This means that it takes close to 2 billion tons of phytoplankton production to support our consumption of swordfish steaks, tuna sashimi, and tuna salads, and all those other great predatory fish that taste so wonderful.

As it happens, these fish are the only ones we harvest from the open ocean. Compare these 2 billion tons of phytoplankton needed to support these fisheries with the 89 billion tons of phytoplankton that the oceans produce annually and the number is quite small.

One more example: the anchoveta, this one from an upwelling. The total catch of this fish alone is three times that of the tuna and similar fish combined, and more in good years. Yet the anchoveta feed on phytoplankton and zooplankton, a much lower trophic level than the tuna. As a consequence, the anchoveta require only one-tenth the amount of phytoplankton production to support it, even though the fishery is much larger. But upwellings occupy an area of much less than 1 percent of the open ocean, and hence the catch is more concentrated. [...]

Pauly and Christensen's work estimated the phytoplankton production that is needed to support the bycatch for the fisheries in different regions in exactly the same way as for the catch itself. [...]

In summary, these were their results.

- Fisheries of the open ocean need about 2 billion tons, and their bycatch 0.23 billion tons of phytoplankton production to support them; combined, this comes to about 2 percent of the 89 billion tons of annual phytoplankton production of the open oceans;
- Fisheries of upwellings need 0.72 billion tons, and their bycatch 0.11 tons; combined, this is 28 percent of the 2.9 billion tons of annual production of upwellings;
- Fisheries of tropical shelves need 1.08 billion tons, and their bycatch 0.1 tons; combined, this is 24 percent of the 5.9 billion tons of annual production;
- Fisheries of non-tropical shelves need 3.08 billion tons, and their by-catch 1.35 tons; combined, this is 35 percent of the 12.5 billion tons of annual production;
- Fisheries of coastal waters and reefs need 0.17 billion tons, and their bycatch 0.05 tons; combined, this is 6 percent of the 4 billion tons of annual production;
- Fisheries of rivers and lakes need 0.3 billion tons; no estimate of bycatch was available; the catch alone is 23 percent of the 1.3 billion tons of annual production.

Averaged across all the oceans, only about 8 percent of the total phytoplankton production goes to support the fisheries we consume. This average is quite misleading, however. Across the open oceans only 2 percent of the productivity supports our fisheries. Yet across the geographically restricted, highly productive parts of the oceans and freshwater lakes, between a quarter and a third of all plant production supports our fisheries.

Pimm's (2001) audit of the Earth is in essence a recalculation of the work of Vitousek et al. (1986) and Pauly and Christensen (1995). Although he uses a different approach and different assumptions, he ends up largely confirming our findings. Why did he bother? Apart from getting a very readable book out of it, he gives the answer:

The Wold According to Pimm

Why did I repeat the calculations? I didn't believe them.... It is nothing personal. I do not believe my estimate(s) any more than I do theirs. Scientists are suspicious, critical nitpickers who are always ready to abuse their friends and colleagues verbally and in print over such differences. This is as it should be. In the arcane, acrimonious debate is a statement of how confident we are about the estimates. Differences in the estimates are to be expected, indeed encouraged. Ecologists calculate different numbers all the time. The key to global accounting is to check numbers by using as many different approaches as possible. When the numbers agree, it is great. When they do not agree, we ask, Why not? In the answer, we hope to find ways of making the numbers more accurate.

I can't say how happy we were to have survived this audit. As for Stuart … he thought it better to give the second edition of *The World According to Pimm* the more sober title *A Scientist Audits the Earth*.

Coverage by the Mass Media

In earlier days, when a researcher published a scientific paper, the extent of the "outreach" initiated was to distribute reprints to colleagues and competitors, and perhaps to the researcher's mother (if this was a first). It was left to journalists to find out that somebody had published something that might have been of interest to the public and, by extension, to administrators and politicians. *Nature* was one of the few scientific journals with its own outreach, through journalists alerted beforehand of potentially interesting findings, on the assumption that they wouldn't publish their articles before the "embargo" was lifted, on the publication date (here the 16th of March 1995).

As it turned out, both Reuters (Fox 1995) and the Associated Press (Ritter 1995) wrote features on our contribution, and they were picked up by newspapers and published the next day under various titles and bylines all over the world. It was a new experience for me to see my work discussed in the media. The fact that it did get so widely reported was a testament to the fact that conflicts over fishing were simmering around the world, and that science would play an increasingly visible role in such disputes. For example, the finding that more of the primary production of the sea is appropriated by humans than previously thought was used by the UK *Guardian* on March 16, in an article titled "New Calculations by Scientists Show There Can Be No Winners in Fight for Dwindling Catches" (Radford 1995) to contextualize then-current issues, foremost the Canadian-Spanish "fish war," a link also suggested in the Reuters feature:

THE (UK) GUARDIAN

New Calculations by Scientists Show There Can Be No Winners in Fight for Dwindling Catches

T. Radford | March 16, 1995

Daniel Pauly, at the University of British Columbia in Vancouver, and a colleague at the International Centre for Living Aquatic Resources in Manila, went back to the marine equivalent of grass roots, and calculated the amount of plankton needed to "feed" the fish that add up to the annual world catch.

They came to the startling conclusion—published in *Nature* this morning—that between 25 and 35 per cent of all the primary production is needed to sustain the current catch in the more valuable fishing grounds. Even if the productivity was averaged over the whole ocean—70 per cent of the globe—the current world catch still added up to 8 per cent of all the plankton the oceans could grow. This is four times greater than all previous estimates.

"That really is quite large," said John Beddington, a professor at the Centre for Environmental Technology at Imperial College, London. "It indicates that there is cause for concern. And the stocks are declining anyway."

The pickings for fishermen over most of the open ocean are slim, but there are a number of rich, traditional grounds. One of them is the Grand Banks off Newfoundland, where Europe and Canada are battling over who has the right to catch how much of which species.

It may seem like a local problem. It isn't. Marine biologists started warning 30 years ago that—contrary to popular legend—there were not plenty more fish in the sea.

Nations—Britain, for instance, and Iceland, and Norway—began fighting over what had once been thought of as a shared resource.

Then fishing boats became more expensively equipped, as fishermen competed for smaller catches. Men who once sailed for Biscay and the Dogger started trawling for hake off Namibia. New, unfamiliar species appeared on the markets. And fishing suddenly became big politics.

The fleets of the world catch rather less than 100 million tonnes a year. Techniques are hugely wasteful: up to 27 million tonnes of fish are thrown overboard every year. Fishing is not even profitable. The UN estimates that the world fishing fleet loses $ 54 billion […] a year. It costs governments something like $ 25 billion in subsidies simply to keep the fleets afloat.

Spain has the largest fishing industry in Europe—19,000 boats, 85,000 men—and keeps 425,000 people busy on shore processing and exporting the catch. At 40 kilograms per head per year, Spaniards eat more fish than any other nation, except for the Koreans and Japanese. But Spain still has to import huge quantities of cod, hake and squid to meet demand.

Somehow, the Spaniards have emerged as the "villains" of the Atlantic

THE (UK) GUARDIAN

war, but the arguments are complicated. Canada protects her fishing interests inside her own 200 mile economic zone. But the once-rich, once free-for-all grounds of the North Atlantic are now "managed" by the North Atlantic Fisheries Organisation, which wants to allot a smaller ration of turbot and halibut to European fishermen, and a larger share to the Canadians.

For once, the European nations are showing remarkable unity: the word from Brussels is that Canada has distorted complicated arguments about fishing limits to disguise the role of its own fishermen in bringing fish stocks close to exhaustion. British fishermen will see this as a bit of a turn up: the view from Cornwall, for instance, is that it is the Spaniards who have been vacuum-cleaning the seabed. But the arrest of the trawler *Estai* in international waters was argued as a defiance of international law. The Canadians see themselves as the defenders of conservation—both of fish stocks and of Newfoundland fishermen.

The lesson of the plankton calculations published today is that this is a war nobody can win. The problem is not political, it is biological. The answers, say scientists, will lie in better management.

Just as evident as the fact that our work would be fodder in these wars was the fact that once our work got into the media, we lost control of the story line. Ritter's feature for the Associated Press dealt fairly with our work, as far as it went, but Ritter also felt compelled to seek different sides of the issue, including in this case Richard Gutting of the National Fisheries Institute, an industry group, who offered a view very much at odds with our findings. Here is the AP story:

ASSOCIATED PRESS

Global Fishing Taking Up Much of Ocean's Algae, Study Says

M. Ritter | March 16, 1995

Commercial fishing takes up a larger share of the ocean's algae than scientists had thought, at the expense of whales, dolphins and other marine life, a new study says.

The competition should be eased by setting up some no-fishing zones along offshore areas called continental shelves, said researcher Daniel Pauly of the University of British Columbia in Vancouver.

He and Villy Christensen analyzed the issue at the International Center for Living Aquatic Resources Management in Manila, Philippines. They present results in Thursday's issue of the journal *Nature*.

Algae are the ocean's ultimate source of food energy, sustaining animals through food chains. The researchers calculated that 8 percent of annual algae production sustains the food chains whose fish are caught commercially. That is nearly four times the previous estimate, they said.

The percentage of algae consumption is much higher for the continental shelves, where Pauly said about 90 percent of the world's catch takes place.

For shelves that are home to such popular species as cod, haddock, herring and flounder, about one-third of the algae essentially work "for us," Pauly said.

The share of algae that sustains commercial fishing is therefore not available to food chains that eventually feed seabirds and large marine mammals like whales and dolphins, he said.

Pauly advocated setting up no-fishing zones on continental shelves.

Richard Gutting Jr., vice president for government relations of the National Fisheries Institute, an industry group, said vast ocean areas have already been set aside as no-fishing zones.

"Here in the United States, the more urgent issue is whether we can continue to feed exploding populations of marine mammals, such as seals and sea lions, which are competing for seafood with human beings," he said.

The last item in this, that "vast ocean areas have already been set aside," was absurdly off the mark—and still would be if uttered today (Wood et al. 2008), as is the bit about marine mammals being our competitors (as if they were taking food from our tables; Kaschner and Pauly 2005; Gerber et al. 2009).

A Large Fermi Solution

One way of looking at our 1995 *Nature* contribution was as a large "Fermi solution." In physics, when there is the need to estimate unknown quantities from limited data, an approach named after the physicist Enrico Fermi is often used. This is often illustrated by his estimation of the number of piano tuners in Chicago, in the absence of specific data. For this, he broke the problem down into parts about which he did have data—number of pianos per household, number of households in Chicago, frequency with which pianos needed to be tuned, etc.—then computed his estimate (von Baeyer 1993).[21] What is obtained with this method is not a definitive number but rather a reasonable estimate, on the basis of which one can then identify critical steps requiring further examination.[22]

Fisheries ecology presents many problems whose solutions require data we do not have, thus forcing us to resort to a large Fermi solution. In this case, we turned to Ecopath models to refine our estimates and get a clearer picture of the critical

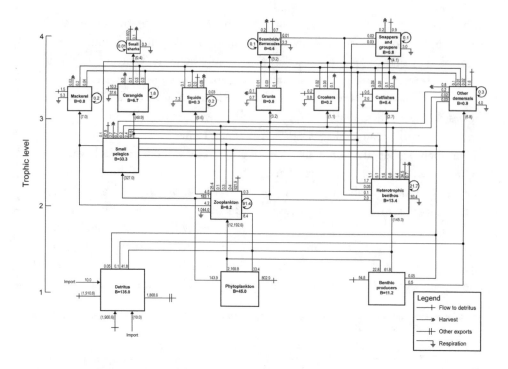

Figure 1.4. Trophic flux model of the northeastern Venezuelan shelf, constructed using the Ecopath approach and software (from Mendoza 1993). Note the reticulated nature of the major fluxes, invalidating the notion of linear food chains, and leading, for the various consumers (boxes), to (fractional) trophic levels that are estimated, i.e., represent model output, rather than assumed input.

parameters.[23] Ecopath models are constructed so that the ecosystem components are first arranged in a number of functional groups. For each of these groups, the production, consumption, and diet are quantified, and a possible model with a set of mutually compatible trophic fluxes is output by the program. An example of such a model is given in Figure 1.4.

Following this approach (which is now widely used and adopted by such agencies as NOAA),[24] we arrived at better estimates of mean transfer efficiency (Figure 1.5). We found that values of 15% and beyond, estimated by earlier authors, would not have been found had actual food webs, rather than hypothetical food chains, been studied. In fact, for reasons discussed in Jarre-Teichmann and Christensen (1998), upwelling systems tend to have much lower than average transfer efficiency, nearly one order of magnitude lower than the mean value 20% assumed by Ryther (1969; see above).

We also confirmed the voice of a dissenter, Moiseev (1994), in his claim that the optimal fishing mortality for a fish stock (i.e., that associated with its "maximum sustainable yield") is much lower than its natural mortality, casting further doubt on

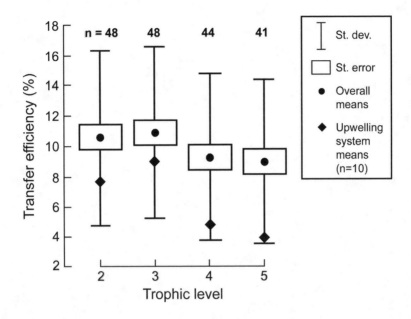

Figure 1.5. Transfer efficiency (%) in 48 models of trophic flows in aquatic ecosystems by trophic level; based on the same sources as Figure 1.3. Note the overall mean of about 10% and the low means for upwelling systems, contradicting earlier assumptions.

the claim that a world catch of 100 million tonnes would be sustainable. (Note that observed catches can be higher for several years; the problem is that they cannot be sustained, a theme followed up on in Chapters 2 and 4.)

In our refinement of the model, we used larger numbers of groups when estimating PPR, which not only has the effect of reducing aliasing bias (Box 1.2), but also leads to robust estimates, wherein an overestimate of one or the other parameter is compensated for, at least in part, by underestimates of other parameters. This effect, which is an essential part of "large Fermi solutions" (see above) has been used by few of the above-cited authors. Only in the FAO study led by Gulland (1970) and in Moiseev (1969) were the world ocean and its resources broken into enough strata for this effect to occur.

One additional feature of such stratification is, obviously, that it allows the identification of problematic or interesting strata. Thus, the information that the PPR required to sustain the world's fisheries is 8% of the world's primary production is not as telling as the information that PPR is very low in open ocean waters (2%) and very high on continental shelves (25–35%), where most fishery catches originate. This, indeed, was the point of our 1995 contribution.[25]

CHAPTER TWO

Fishing Down the Food Web

Another Summer in Manila

In 1998, three years after the outburst of work leading to the contribution presented in Chapter 1, I found myself once again summering in Manila. ICLARM had much changed since I had worked there full time, and indeed, the organization was soon to change its name, and decamp to Malaysia.[1] But it still had then small islands of productivity, notably Rainer Froese's FishBase and Villy Christensen's Ecopath projects, which made it possible to continue working with the same productive gang, this time on change in species composition as a result of fishing.

There was evidence well before the 1990s that fishing alters the species composition of marine ecosystems. For example, researchers found that the abundance of large fish in the Gulf of Thailand tended to decline faster than total biomass (Tiews et al. 1967; Pauly 1979a; Beddington and May 1982), and another study showed that large skates had disappeared from the Irish Sea (Brander 1981). However, as illustrated by the contributions in May (1984), we lacked an easy-to-apply measure that could have been used to compare such impact between ecosystems, scale up from local data sets, and allow for the identification of global trends.

Working on the primary production requirements of various fisheries, and developing and teaching food web–based approaches for ecosystem modeling (Chapter 1) had made me aware of the usefulness of trophic levels (TL) as tools with which to organize knowledge on aquatic ecosystems, notwithstanding earlier rejections of the trophic level concept (Rigler 1975; Cousins 1985).[2]

Indeed, a growing body of work with Ecopath models of freshwater and marine ecosystems persuaded me that trophic levels are essential to understanding the health of aquatic ecosystems. Put simply, an ecosystem whose average trophic level was declining was almost certainly declining in health, and perhaps at risk of collapse. Therefore, I had pushed for the earliest version of FishBase (see Box 2.1) to include not only diet composition data for all fish species whose food and feeding habits had been published for, but also the trophic level corresponding to each set of diet composition data, to be computed automatically using a routine developed for the purpose (Pauly and Christensen 2000).[3]

As a result, I had seen hundreds of estimates of trophic levels, for a wide range of aquatic animals, and had even initiated a review of their values in marine mammals (later published as Pauly et al. 1998c) when, for a course held at the University of British Columbia in the first months of 1997, I asked one of my graduate students, Ms. Johanne Dalsgaard, to include a plot of the mean trophic levels of landings in a term paper describing, based on FAO fisheries statistics, the fisheries of FAO Area 27 (see Figure 1.1), which comprises the northeast Atlantic Ocean, an area in which species are relatively well identified (or "disaggregated") in landing data.

The resulting plot of the mean trophic level of the fishes caught by these fisheries from 1950 to the mid-1990s was nearly linear, with a negative slope, indicating that European fisheries catches are increasingly composed of fishes from the lower part of marine food webs (i.e., prey, or forage fishes), notwithstanding the effort of the International Council for the Exploration of the Sea (ICES) to ensure the sustainability of these fisheries.

Ms. (now Dr.) Dalsgaard got a good grade for her term paper, and I forgot all about it. But then, a few weeks later, I was back in Manila, as part of the transpacific commute mentioned in Chapter 1, discussing indicators of ecosystem status with Villy Christensen, Rainer Froese, and other friends and colleagues. More precisely, it was in a little greasy spoon, the Singapura Restaurant, where I was having (low-trophic-level) *sambal* squids with rice. What happened is that I suddenly realized that the plot by Johanne Dalsgaard could be easily expanded to cover the whole world. In fact, the contribution to be presented below exploded into existence in my head, and "all" that was left was … to do the work.

Villy Christensen saw the point right away, notably because he had been asked by FAO to write a review of how to manage fisheries by focusing on the interaction between top predators and their prey, as a background document for a fisheries conference in Kyoto, Japan.[4] This paper, in a section on "fishing down the food web," explored the relationship between the level of fisheries catches and the trophic levels of species that the fisheries target for 36 Ecopath models of aquatic ecosystems. The relationship indicated a strong negative correlation between catch and trophic level and that a decrease of one trophic level in the target fishery results in a yield 8.3 times higher—surprisingly close to the 10 times that could be expected if trophic transfer efficiency were 10% (Christensen 1996). This was a nice confirmation, incidentally, of the arguments for the mean value TE of 10% used in Chapter 1.[5]

Rainer Froese, the leader of the FishBase project at ICLARM, also immediately saw the point. Moreover, he had previously made the FAO catch (actually "landing")

Box 2.1
What is FishBase?

FishBase is an online database, available at www.fishbase.org, with key data on the biology of all fishes. There are over 30,000 species of marine and freshwater bony and cartilaginous fishes (i.e., fish in a narrower sense, not as defined in Box 1.1) and all are covered.

FishBase was initially conceived as a tool to assist in the management of fisheries in the tropics, and hence the early emphasis on the distribution and population dynamics of tropical species. However, we quickly realized that the distinction between commercial and other fishes, or tropical and other fishes, was arbitrary, and we grew FishBase to accommodate all species of fish in the world. (This is a moving target, incidentally, as a few hundred species are newly described every year).

Essentially, we endeavor to cover all aspects of a species' biology and interactions with humans. On the biological side, this involves describing their morphology (including photos and drawings), physiology and genetics, and their food and feeding, reproduction, habitats, and general ecology. We also cover, obviously, their taxonomy and classification using William Eschmeyer's monumental *Catalog of Fishes* (Eschmeyer 1998) as backbone and, less obviously, their common names (200,000+ names in over 200 languages). The human dimension is also present in the distribution by countries, and it allows various national species lists to be output, notably, on their threat status. All information provided is linked to its source, usually a scientific article, and to the collaborator who provided that piece of information.

FishBase was developed at the International Center for Living Aquatic Resources Management (ICLARM, now WorldFish Center) starting in the late 1980s, in collaboration with the Food and Agriculture Organization (FAO) of the United Nations and many other institutional partners and individual collaborators. It was funded mainly through sequential grants from the European Commission.

FishBase, which is now run by an international consortium of nine research institutions on three continents and which offers information through interfaces in numerous languages and scripts, has several mirror sites (e.g., fishbase.de, fishbase.fr), which enable it to accommodate very high traffic (over 30 million hits per month) caused by around one million visitors monthly.

The Web site (www.fishbase.org) provides a downloadable manual, also available in hard copy (Froese and Pauly 1997, 1998, 1999, 2000), which presents, among other things, the rationale for the choice of information that is encoded in FishBase.

statistics accessible through FishBase and programmed the FishBase routine that computed regional catch compositions over time. All that was needed, additionally, was for the FishBase programmer, Ms. A.G. Laborte, to link these time series with the trophic level estimates that Francisco Torres Jr. had assigned to each of the 1200+ taxa in the FAO statistics, based on 200+ independent trophic level estimates from Ecopath. This yielded a routine that called on the global FAO data set (incorporated into FishBase) to compute, at the press of a button, country- and region-specific time series of mean trophic level, of primary production required, and of other ecosystem status indicators (see section on FAO Statistics, p. 122–128 in Froese and Pauly 2000).

Our contribution literally wrote itself. It was finished in a few days, and we submitted it to *Science*.

I can't resist mentioning here that one of the referees felt that the "data" we presented (i.e., time series of trophic levels) were far more interesting than the interpretation we provided. I did not mind that, as this did not preclude a recommendation to publish. Here is how *Science* introduced "fishing down marine food webs":

SCIENCE

Fishing Too Deep

What is the future for fisheries worldwide? Bleak, according to an analysis by Pauly et al. [...]. From detailed assignations of trophic level for 220 marine and freshwater species, they have assessed patterns in global and regional fish-catches in long-term data. For virtually all fisheries, the average trophic level of landed catches is falling: short-lived, low trophic level invertebrates, and planktivores are replacing long-lived, high trophic level species. The apparent unsustainability of fisheries represents an urgent challenge to ocean management.

Our findings were also preceded by their description in a Research News piece in *Science*. (Williams 1998):

SCIENCE

Marine Ecology:
Overfishing Disrupts Entire Ecosystems

In the face of declining fish stocks, the managers of many of the world's fisheries have been forced to take often drastic measures to prevent total collapse. These include, for example, a complete ban on fishing the Grand Banks off Newfoundland and quotas that limit takes, such as those now imposed on fishing vessels in European Union waters. But a new analysis of global fish catches over the past 45 years, which appears on page 860, suggests that even more drastic action is urgently needed.

The study—conducted by Daniel Pauly and Johanne Dalsgaard of the University of British Columbia in Vancouver and colleagues at the International Center for Living Aquatic Resources Management in Makati, the Philippines—concludes that humans are inexorably fishing down marine food webs as larger and more commercially valuable species disappear, creating impoverished, less valuable ecosystems. Complete fishing bans currently apply to less than 1% of the world's fishing grounds, but fisheries experts say the findings of this new study indicate that more such protected areas must be created if there is to be any chance of salvaging vanishing ecosystems. "Most researchers work at the fishery or species level, but this study looks at the global picture and reveals just how unsustainable our exploitation of marine resources is. It's a wake-up call," says marine researcher Elliott Norse, president of the Marine Conservation Biology Institute in Redmond, Washington.

To come to this conclusion, Pauly, Dalsgaard, and their colleagues first used an analysis of the diet of 220 key species to assign to each species of catch a trophic level, a rating describing its location in the food chain. Trophic level 1 comprises the primary photosynthetic plankton, while a top predator, such as the snappers inhabiting the continental shelf off Mexico's Yucatán Peninsula, gets a rating of 4.6.

Then, the team analyzed data collected by the United Nations Food and Agriculture Organization on catches in the world's major fisheries from 1950 to 1994 to determine whether the trophic levels had changed with time. This showed that there had been a gradual shift from long-lived, high-trophic-level fish (such as cod and haddock) to low-trophic-level invertebrates and plankton-feeding fish (such as anchovy). Overall, the researchers found a steady mean decline of about 0.1 trophic levels per decade in the worldwide catches. What's more, Pauly says, "this is probably an underestimate, as catch measurements from the tropics are poorly recorded."

The results also indicate that the quantities, as well as the quality, of the catches are decreasing. At first, skimming off the top of the food chain and then moving down to lower trophic levels can lead to increased catch sizes, because top predators require a large reservoir of prey to sustain them. But the new research shows that, in most instances, when the top predators are removed, catches stagnated or declined, apparently because the populations of the predators' competitors for food expand. "The Black Sea provides a good example," Pauly says. "There's been a huge increase in jellyfish as their economically valuable competitors have been removed."

As a result of this overfishing, the number of main fisheries in the Black Sea has fallen from 26 in the 1970s to five now, says Norse. "Present fishing policy is unsustainable. The food-web structure is changing," says Pauly. "At least 60% of the world's 200 most commercially valuable species are overfished or fished to the limit," says Claude Martin, director-general of the World Wide Fund for Nature.

Pauly argues that there is an urgent need to create protected areas, where fishing is not allowed. Although other measures, such as quotas, limiting fishing time at sea, changing fishing gear, and controlling pollution are crucial, they are difficult to

SCIENCE

implement quickly and control, he says. And there is growing evidence that protected areas can be highly effective in restoring and maintaining marine ecosystems. Such areas on the Georges Bank off Massachusetts were created only in 1994, but researchers are already finding an increase in the size and spawning populations of key fish species, as well as a rapid increase in the bottom-dwelling scallop population, says a spokesperson for the National Marine Fisheries Service in Woods Hole, Massachusetts.

Even tiny protected areas can be very effective in some regions. Callum Roberts of the University of York in the United Kingdom says reserves of just a few hectares on tropical coral reefs have boosted fish stocks and helped maintain long-lived large predators [Figure 2.1]. The fishing industry is also now beginning to back this policy. In the United Kingdom, the industry now backs plans for no-fishing areas as a key way to develop the European Union's fishing policy in the face of declining stocks, says Roberts. "At the very least, they can offer quick and simple protection while the complexity of long-term sustainable fishing policies are developed," says Norse.

But Pauly's results have set a clock ticking on the development of such policies. "In 30 to 40 years, our fisheries could have moved down another 0.5 of a trophic level in overall catch, which is an enormous change," he says. "If things go unchecked, we might end up with a marine junkyard dominated by plankton."

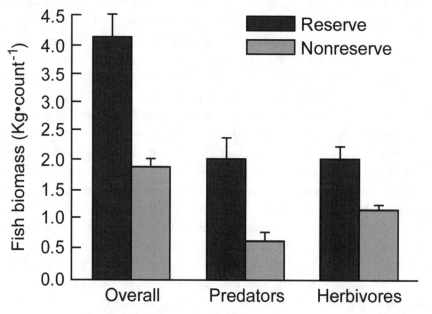

Figure 2.1. Bouncing back: Fish stocks recovered two years after a small reserve was set up off St. Lucia (adapted from Roberts and Polunin 1993).

Here is the full text of our report:

SCIENCE

Fishing Down Marine Food Webs[6]

D. Pauly et al. | Vol. 279 No. 5352

The mean trophic level of the species groups reported in Food and Agricultural Organization global fisheries statistics declined from 1950 to 1994. This reflects a gradual transition in landings from long-lived, high trophic level, piscivorous bottom fish toward short-lived, low trophic level invertebrates and planktivorous pelagic fish. This effect, also found to be occurring in inland fisheries, is most pronounced in the Northern Hemisphere. Fishing down food webs (that is, at lower trophic levels) leads at first to increasing catches, then to a phase transition associated with stagnating or declining catches. These results indicate that present exploitation patterns are unsustainable.

Exploitation of the ocean for fish and marine invertebrates, both wholesome and valuable products, ought to be a prosperous sector, given that capture fisheries—in contrast to agriculture and aquaculture—reap harvests that did not need to be sown. Yet marine fisheries are in a global crisis, mainly due to open access policies and subsidy-driven over-capitalization (Garcia and Newton 1997). It may be argued, however, that the global crisis is mainly one of economics or of governance, whereas the global resource base itself fluctuates naturally. Contradicting this more optimistic view, we show here that landings from global fisheries have shifted in the last 45 years from large piscivorous fishes toward smaller invertebrates and planktivorous fishes, especially in the Northern Hemisphere. This may imply major changes in the structure of marine food webs.

Two data sets were used. The first has estimates of trophic levels for 220 different species or groups of fish and invertebrates, covering all statistical categories included in the official Food and Agricultural Organization (FAO) landings statistics (FAO 1996). We obtained these estimates from 60 published mass-balance trophic models[7] that covered all major aquatic ecosystem types (Christensen and Pauly 1993b; Pauly and Christensen 1993; Christensen 1995a; Pauly and Christensen 1995; see also note[a]). The models were constructed with the Ecopath software (Christensen and Pauly 1992a) and local data that included detailed diet compositions (note[b]).

[a] The bulk of the 60 published models are documented in Christensen and Pauly (1993b) [and] Pauly and Christensen (1993, 1995). References to the remaining models are given in Froese and Pauly (1997).

[b] The documentation of the Ecopath models in Christensen (1995a) and [the references listed] in the note above includes sources of diet compositions of all consumer groups in each ecosystem. These diet compositions are rendered mutually compatible when mass-balance within each model is established.

In such models, fractional trophic levels[c] are estimated values, based on the diet compositions of all ecosystem components rather than assumed values; hence, their precision and accuracy are much higher than for the integer trophic level values used in earlier global studies (Ryther 1969). The 220 trophic levels derived from these 60 Ecopath applications range from a definitional value of 1 for primary producers and detritus to 4.6 (+/− 0.32) for snappers (family Lutjanidae) on the shelf of Yucatan, Mexico (note[d]). The second data set we used comprises FAO global statistics (FAO 1996) of fisheries landings for the years from 1950 to 1994, which are based on reports submitted annually by FAO member countries and other states and were recently used for reassessing world fisheries potential (Grainger and Garcia 1996). By combining these data sets we could estimate the mean trophic level of landings, presented here as time series by different groupings of all FAO statistical areas and for the world (note[e]).

For all marine areas, the trend over the past 45 years has been a decline in the mean trophic level of the fisheries landings, from slightly more than 3.3 in the early 1950s to less than 3.1 in 1994 (Figure 2.2A). A dip in the 1960s and early 1970s occurred because of extremely large catches [$>12 \times 10^6$ metric tons (t) per year] of Peruvian anchoveta with a low trophic level (Jarre et al. 1991) of 2.2 ($\pm\,0.42$). Since the collapse of the Peruvian anchoveta fishery in 1972–1973, the global trend in the trophic level of marine fisheries landings has been one of steady decline. Fisheries in inland waters exhibit, on the global level, a similar trend as for the marine areas Figure 2.2B): A clear decline in average trophic level is apparent from the early 1970s, in parallel to, and about 0.3 units below, those of marine catches. The previous plateau, from 1950 to 1975, is due to insufficiently detailed fishery statistics for the earlier decades (Grainger and Garcia 1996).

[c] As initially proposed by Odum and Heald (1975).

[d] All trophic level estimates [were] fully documented on the home pages of the Fisheries Centre, University of British Columbia [...]. They can also be found on the FishBase 97 CD-ROM [and its annual updates to 2000 (Froese and Pauly 1997, 1998, 1999, 2000)], which also presents the references to the 60 published Ecopath applications. FishBase 97 also includes the FAO statistics, so [the above figures] can be reproduced straightforwardly. To estimate the standard error (SE) we used the square root of the variance of the estimate of trophic level, in agreement with Pimm (1982), who defined an omnivore as "a species which feeds on more than one trophic level." Thus, our estimates of SE do not necessarily express uncertainty about the exact values of trophic level estimates; rather, they reflect levels of omnivory. We do not present SE for the trophic levels of fisheries landings, as fisheries are inherently "omnivorous."

[e] Mean trophic level TL_k for year k, is estimated by multiplying the landings (Y_{ik}) by the trophic level of the individual species groups i, then taking a weighed mean, that is, $TL_k = \Sigma_i\, TL_i \cdot Y_{ik} \,/\, \Sigma_i\, Y_{ik}$.

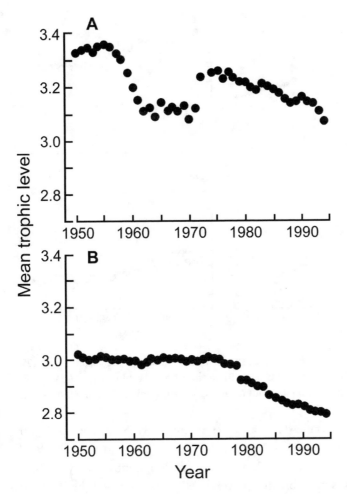

Figure 2.2. Global trends of mean trophic level of fisheries landings, 1950–1994. (A) Marine areas; (B) inland areas.

In northern temperate areas where the fisheries are most developed, the mean trophic level of the landings has declined steadily over the last two decades. In the North Pacific (FAO areas 61 and 67; Figure 2.3A), trophic levels peaked in the early 1970s and have since then decreased rapidly in spite of the recent increase in landings of Alaska pollock, *Theragra chalcogramma*, which has a relatively high trophic level of 3.8 (± 0.24). In the Northwest Atlantic (FAO areas 21 and 31; Figure 2.3B), the fisheries were initially dominated by planktivorous menhaden, *Brevoortia* spp., and other small pelagics at low trophic levels. As their landings decreased, the average trophic level of the fishery initially increased, then in the 1970s it reversed to a steep decline. Similar declines are apparent throughout the time series for the Northeast Atlantic (FAO area 27; Figure 2.3C) and the Mediterranean (FAO area 37; Figure 2.3D), although the latter system operates at altogether lower trophic levels.

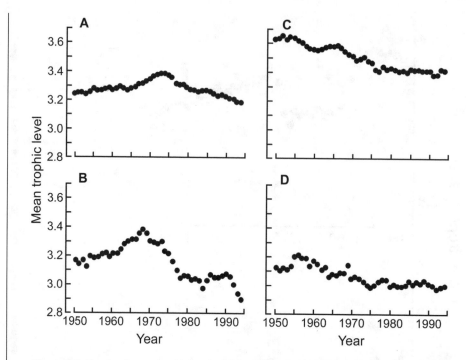

Figure 2.3. Trends of mean trophic level of fisheries landings in northern temperate areas, 1950–1994. (A) North Pacific (FAO Areas 61 and 67); (B) northwest and western central Atlantic (FAO Areas 21 and 31); (C) northeast Atlantic (FAO Area 27); and (D) Mediterranean (FAO Area 37).

The Central Eastern Pacific (FAO area 77; Figure 2.4A), Southern and Central Eastern Atlantic (FAO areas 41, 47, and 34; Figure 2.4B), and the Indo-Pacific (FAO areas 51, 57, and 71; Figure 2.4C) show no clear trends over time. In the southern Atlantic this is probably due to the development of new fisheries, for example, on the Patagonian shelf, which tends to mask declines of trophic levels in more developed fisheries. In the Indo-Pacific area, the apparent stability is certainly due to inadequacies of the statistics, because numerous accounts exist that document species shifts similar to those that occurred in northern temperate areas (e.g., Beddington and May 1982; Dalzell and Pauly 1990; Silvestre and Pauly 1997b).

The South Pacific areas (FAO areas 81 and 87; Figure 2.5A) are interesting in that they display wide-amplitude fluctuations of trophic levels, reflecting the growth in the mid-1950s of a huge industrial fishery for Peruvian anchoveta. Subsequent to the anchoveta fishery collapse, an offshore fishery developed for horse mackerel, *Trachurus murphyi*, which has a higher trophic level (3.3 ± 0.21) and whose range extends west toward New Zealand (Parrish 1989). Antarctica (FAO areas 48, 58, and 88; Figure 2.5B) also exhibits high-amplitude variation of mean trophic levels, from a high of 3.4, due to a fishery that quickly depleted local accumulations of

SCIENCE

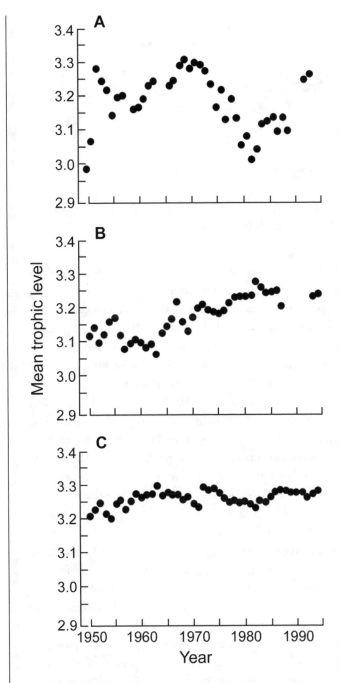

Figure 2.4. Trends of mean trophic levels of fisheries landings in the intertropical belt and adjacent waters. (A) Central eastern Pacific (FAO Area 77); (B) southwest, central eastern, and southeast Atlantic (FAO Areas 41, 34, and 47); and (C) Indo (west) Pacific (FAO Areas 51, 57, and 71).

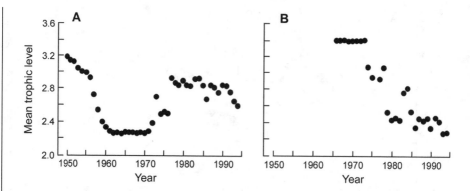

Figure 2.5. High-amplitude changes of mean trophic levels in fisheries landings. (A) South Pacific (FAO Areas 81 and 87); (B) Antarctica (FAO Areas 48, 58, and 88).

bony fishes, to a low of 2.3, due to *Euphausia superba* (trophic level 2.2 ± 0.40), a large krill species that dominated the more recent catches.

The gross features of the plots in Figures 2.2 through 2.5, while consistent with previous knowledge of the dynamics of major stocks, may provide new insights on the effect of fisheries on ecosystems. Further interpretation of the observed trends is facilitated by plotting mean trophic levels against catches. For example, the four systems in Figure 2.6 illustrate patterns different from the monotonous increase of catch that may be expected when fishing down food webs (Christensen 1996). Each of the four systems in Figure 2.6 has a signature marked by abrupt phase shifts. For three of the examples, the highest landings are not associated with the lowest trophic levels, as the fishing-down-the-food-web theory would predict. Instead, the time series tend to bend backward. The exception (where landings continue to increase as trophic levels decline) is the Southern Pacific (Figure 2.6C), where the westward expansion of horse mackerel fisheries is still the dominant feature, thus masking more local effects.

The backward-bending feature of the plots of trophic levels versus landings, which also occurs in areas other than those in Figure 2.6, may be due to a combination of the following: (i) artifacts due to the data, methods, and assumptions used; (ii) large and increasing catches that are not reported to FAO; (iii) massive discarding of bycatches (Alverson et al. 1994) consisting predominantly of fish with low trophic levels; (iv) reduced catchability as a result of a decreasing average size of exploitable organisms; and (v) fisheries-induced changes in the food webs from which the landings were extracted. Regarding item (i), the quality of the official landing statistics we used may be seen as a major impediment for analyses of the sort presented here. We know that considerable under- and misreporting occur (Alverson et al. 1994). However, for our analysis, the overall accuracy of the landings is not of major

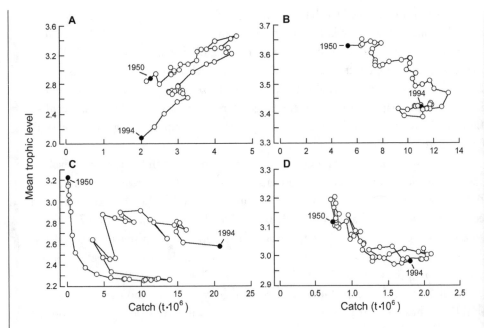

Figure 2.6. Plots of mean trophic levels versus the fisheries landings (in millions of metric tons) in four marine regions, illustrating typical backward-bending signatures (note variable ordinate and abcissa scales). (A) Northwest Atlantic (FAO Area 21); (B) northeast Atlantic (FAO Area 27); (C) southeast Pacific (FAO Area 87); (D) Mediterranean (FAO Area 37).

importance, if the trends are unbiased. Anatomical and functional considerations support our assumption that the trophic levels of fish are conservative attributes and that they cannot change much over time, even when ecosystem structure changes (note[f]). Moreover, the increase of young fish as a proportion of landings in a given species that result from increasing fishing pressure would strengthen the reported trends, because the young of piscivorous species tend to be zooplanktivorous (Robb and Hislop 1980; Longhurst and Pauly 1987) and thus have lower trophic levels than the adults. Items (ii) and (iii) may be more important for the overall explanation. Thus, for the Northeast Atlantic, the estimated (Alverson et al. 1994) discard of 3.7×10^6 t year^{-1} of bycatch would straighten out the backward-bending curve of Figure 2.6B.

[f] We refer here to gill rakers, whose spacing determines the size of organisms that may be filtered, the length of the alimentary canal, which determines what may be digested, or the caudal fin aspect ratio, which determines attack speed, and hence, which prey organism may be consumed. See de Groot (1971) for an example for flatfishes (order Pleuronectiformes).

Item (iv) is due to the fact that trophic levels of aquatic organisms are positively[g] related to size. Thus, the relation between trophic level and catch will always break down as catches increase: There is a lower size limit for what can be caught and marketed, and zooplankton is not going to be reaching our dinner plates in the foreseeable future. Low catchability due to small size or extreme dilution ($<1\,g\cdot m^{-3}$) is, similarly, a major reason why the huge global biomass ($\approx 10^9$ t) of lanternfish (family Myctophidae) and other mesopelagics (Gjøsaeter and Kawaguchi 1980) will continue to remain latent resources.

If we assume that fisheries tend to switch from species with high trophic levels to species with low trophic levels in response to changes of their relative abundances, then the backward-bending curves in Figure 2.6 may be also due to changes in ecosystem structure, that is, item (v). In the North Sea, Norway pout, *Trisopterus esmarkii*, serves as a food source for most of the important fish species used for human consumption, such as cod or saithe. Norway pout is also the most important predator on euphausiids (krill) in the North Sea (Christensen 1995b).

We must therefore expect that a directed fishery on this small gadoid (landings in the Northeast Atlantic are about 3×10^5 t year^{-1}) will have a positive effect on the euphausiids, which in turn prey on copepods, a much more important food source for commercial fish species than euphausiids. Hence, fishing for Norway pout may have a cascading effect, leading to a build-up of non-utilized euphausiids. Triangles such as the one involving Norway pout, euphausiids, and copepods, and which may have a major effect on ecosystem stability, are increasingly being integrated in ecological theory (Pimm 1982), especially in fisheries biology (Jones 1982).

Globally, trophic levels of fisheries landings appear to have declined in recent decades at a rate of about 0.1 per decade, without the landings themselves increasing substantially. It is likely that continuation of present trends will lead to widespread fisheries collapses and to more backward-bending curves such as in Figure 2.6, whether or not they are due to a relaxation of top-down control (Power 1992). Therefore, we consider estimations of global potentials based on extrapolation of present trends or explicitly incorporating fishing-down-the-food-web strategies to

---

[g] Contributions in Christensen and Pauly (1993a) document the strong correlation between size and trophic level in aquatic ecosystems, a case also made for the North Sea by Rice and Gislason (1996). [Also,] we take this opportunity to (i) correct a slip of the pen in our report [Pauly et al. 1998a, p. 862], in which we wrote of an "inverse" relationship between trophic level and length: the two are positively correlated [Figure A2.2]; and (ii) point out that trophic-level estimates based on diet composition data analyzed with Ecopath correlate closely with trophic-level estimates based on stable isotope ratios, notwithstanding size effects (Kline and Pauly 1998). [The second part of this note was originally part of our rejoinder to Caddy et al., now Appendix 2.]

be highly questionable. Also, we suggest that in the next decades, fisheries management will have to emphasize the rebuilding of fish populations embedded within functional food webs, within large "no-take" marine protected areas (Alcala and Russ 1990; Carr and Reed 1993; Dugan and Davis 1993; Roberts and Polunin 1993).

FAO's Comments and a Rejoinder

By far the most interesting critique of "fishing down marine food webs" was by a group of FAO staff (Caddy et al. 1998b), who suggested that the FAO catch statistics used in our contribution were not detailed and reliable enough to support the inference drawn from them. (I don't dare to think what they would have written, had we written such a thing about the FAO statistics!) This critique is presented below in its entirety and is followed by detailed comments, while our first rejoinder (Pauly et al. 1998b) is presented in full as Appendix 2.

How Pervasive Is "Fishing Down Marine Food Webs"[8]
J.F. Caddy et al. | Vol. 282

In their report (Pauly et al. 1998a), and in an earlier [contribution] (Pauly and Christensen 1995), D. Pauly et al. draw global conclusions about the effects of fishing on world fish stocks with the use of research data fitted to Ecopath models at different sites through the world's oceans, integrated with data on global fishery landings collected by the Food and Agricultural Organization of the United Nations (FAO). Although Pauly et al. are to be congratulated for giving this important issue high profile, they greatly oversimplify the situation with their hypothesis and may have misinterpreted the FAO statistics. We do not disagree that a general decline in mean trophic level of marine landings is likely to have occurred in many regions, but we are not convinced that the explanation is solely a result of "fishing down the food web" or that the analysis of the FAO data, as undertaken by Pauly et al., substantiates such a thesis.

Four considerations significantly qualify the evidence of a "fishing down the food web" phenomenon.

(i) Taxonomic resolution. Assigning a trophic level arguably requires knowing at least the related genus or even the species and its age, since trophic level may change by as much as three points from birth to maturity for some top predators. Although the FAO fishery landings data (FAO 1997b) used in their analysis integrate the best estimates by countries, regional fishery organizations, and FAO of the species composition of annual production, it is to be regretted that over 30% of all marine landings cannot be identified to the species level, and about 20% cannot even be assigned to the level of Family (this rises to about 60% for inland capture fishery production). As a consequence, the small drop in mean trophic level they report, from 3.3 in the early 1950s to 3.1 in 1994, appears difficult to substantiate statistically, and its sensitivity to the assumptions necessarily made in allocating coarse data to trophic levels has not been described.

(ii) Landing data as ecosystem indicators. The analysis assumes that changes in mean trophic level in the landings reflect changes in the ecosystem, with the use of annual quantities of landings (excluding discarded catch) as abundance indicators. However, the composition of historical landings has been affected by a number of phenomena that are not simply related to increased fishing pressure (for example, natural oscillations in abundance, changes in fishing technology) and that are likely to have seriously influenced mean trophic levels in the landings.

Overfishing has seriously affected top predators, and this has already been raised in FAO reports (FAO 1992). Peaks in predatory demersal fish production and subsequent declines have been registered that differ in timing regionally and among different habitat types (Grainger and Garcia 1996; Garcia and Newton 1997; Caddy et al. 1998a). Comparing regional landings of demersal fish (generally, long-lived species high in the food chain) with short-lived species such as squid (Caddy and Rodhouse 1998) also reveals trophodynamic effects quite clearly. There seem to be few other hypotheses to account for declines in landings of top predators than overfishing.

The situation is not the same for species lower in the food chain, where natural medium-term fluctuations of the small pelagic species abundance are likely to quantitatively mask effects that result from declining top predators on the mean trophic level. In addition, long-term changes in strategies of fishing such species add to the difficulty of documenting global trends through a "mean trophic level" for the ecosystem as a whole. This task requires detailed knowledge of local fisheries in order to extrapolate safely from "trophic level of landings" to "trophic level of ecosystems."

Outside the north boreal area, and except for a few very large oscillating stocks (for example, Peruvian anchoveta), small pelagics were rarely subjected to major exploitation in the 1960s and 1970s because of lower market prices and because technologies for handling and processing the catch were not yet fully developed. During the last two decades, these species have seen a significant increase in their exploitation resulting from the spread of new technologies. One could argue that this increase is indeed due to increased abundance of pelagics resulting from depletion of their predators, but this remains conjecture. The fact is that the interest of industries for small pelagic fish increased, leading to higher landings of these species and shifts in the composition of global landings. A shift in global fishing strategies could be confused with a "fishing down the food web phenomenon."

(iii) Aquaculture development. FAO landing statistics have traditionally included both capture and aquaculture production, but work is under way to disaggregate them into the two separate components. This has so far been completed for years since 1984. As a rough check, mean trophic levels for species groups as reported in Pauly and Christensen (1995) were applied to marine landings of the corresponding species groups to calculate the overall mean trophic level of total production (capture fishery plus aquaculture production) since 1950 and capture fisheries and aquaculture since 1984. In contrast to the decline in mean trophic level reported by Pauly et al. (1998a), for marine waters the mean trophic level for capture fishery landings has remained stable since 1984 (Figure 2.7)[9] at a level similar to that of total production in the early decades when marine aquaculture was insignificant. The decline in mean trophic level in the total production series is entirely a result of the increasing contribution of aquaculture to total production (from 8% in 1984 to 17% in 1996) and the fact that species cultured in the sea (mainly shellfish) have an average trophic level about half that of capture fishery landings (Figure 2.7). That the results of Pauly et al. (1998a) for all marine waters more closely resemble the trend of total production rather than that of capture fishery landings in Figure 2.7 suggests that aquaculture production may not have been fully excluded from their analysis.

Capture fishery data for inland waters are lacking in taxonomic definition: about 60% are not even assigned to the Family level, and so trophic analysis is not possible. Even if there is a decline in the mean trophic level for inland capture fisheries, it need not necessarily reflect "fishing down the food chain," because there are complications such as large-scale stock-enhancement practices (for example, stocking to the wild, fertilizing reservoirs) as well as pollution that will affect species composition.

(iv) Eutrophication of coastal areas. Accumulating evidence from coastal and semi enclosed seas suggests that land-based runoff, by increased primary productivity along coastlines, may have exerted a "bottom up" effect in increasing abundance of planktivores, thus lowering mean trophic level. This effect is most easily documented where anthropogenic eutrophication has occurred. In the Black Sea, hypoxic effects have decimated demersal species, again decreasing mean trophic level without necessarily implying "fishing down the food web." Similar examples may be cited from the Baltic (Hansson and Rudstam 1990), Black Sea (Zaitsev 1992), Mediterranean (Caddy et al. 1995), and Seto Inland Sea (Tatara 1991). Hypoxia

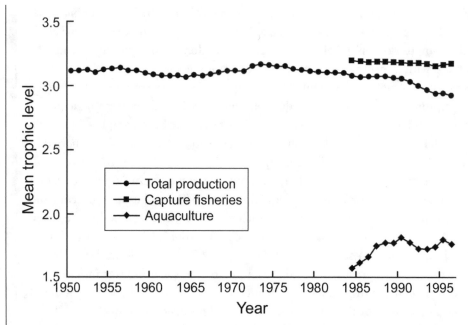

Figure 2.7. Trends in mean trophic level of landings from marine waters (reproduced from Figure 1 in Caddy et al. 1998b).

has even been recently documented as a serious problem in open-sea areas such as the Gulf of Mexico (Rabalais et al. 1996). As a diffuse and general phenomenon, eutrophication is a strong potential source of modification of the ratio between demersal and pelagic fish and between predator and prey abundances that could also be confused with "fishing down the food web."

All these points imply that, even if the mean trophic level of landings was higher earlier on (which in our view is not proven), this does not necessarily reflect "fishing down the food web," because overall landings have increased substantially in recent decades, contrary to what was stated in Pauly et al. (1998a).[10] We concur with Pauly et al. that the mean trophic level for most marine fish species rises with age, which adds another level of uncertainty to their analysis. For example, a bluefin tuna may rise by three trophic levels or more during its life history, as may other large predators that are planktivorous in the larval and post larval stages. This variation makes assigning a single trophic level to a species, which is the practice in Ecopath models, a hazardous procedure. As noted by Pauly et al., the model may actually have underestimated the decline in mean trophic level as the mean age and trophic level of a species has declined with increasing fishing intensity.

We do not mean to imply that "fishing down the food chain" is not a major cause of changes to fish communities worldwide, but the situation of marine fisheries is complex (Botsford et al. 1997), and shows wide regional variation. Oversimplifying a key issue like this could inhibit local research on human impacts on marine food chains that should not be confined to impacts of the fishing industry. Local analyses of food webs using methods promoted by Pauly et al. should be combined with local knowledge of fisheries and research data that take into account possible causal hypotheses. Identification of specific effects of human activities could then lead to locally appropriate management solutions.

There were several factual errors in this response with which we took issue,[11] and which we addressed in our initial published response (Appendix 2), but which for our purposes here are unimportant. It was clear that Caddy et al.'s critique raised a set of questions that required deeper exploration, and in fact, those four points perfectly articulated a research program that I pursued in the years following the publication of "fishing down marine food webs." As it turned out, each of Caddy et al.'s points reinforced our original conclusions. In effect, they became "judo arguments" (Asimov 1977), whose detailed examination strengthened the case that we were "fishing down the food web."

Taxonomic Resolution

The issue of "taxonomic resolution," as viewed by Caddy et al., has two parts. The first, they suggested, is that fish, in the course of their ontogeny, change their food organisms, and hence their trophic levels. For example, the trophic level of fishes in the eastern Mediterranean has been shown to vary seasonally (Karachle and Stergiou 2006). However, this is not likely to be relevant here, as the fishing-down analyses are conducted at the scale of decades. More pertinent here is the fact that most fish (except herbivores, which contribute minuscule amounts to global fisheries catches) have a lower trophic level when they are small and young than when they are large and old (see Figure A2.2, panel B).

Thus, as fishing mortality increases through time (and it does: this is the whole point about how overfishing induces changes in species composition and fishing down), the proportion within a species of large (and hence high-trophic-level) fish will decline, and that of small (low-trophic-level) fish will increase. This can also be expressed through agonizing equations, and the alert reader can find such in Pauly et al. (2001a), where we describe an age-structured model applied to northern cod off eastern Canada, and in Pauly and Palomares (2001), where a generic length-structured model is presented. The result: not considering the relationship between ontogeny and trophic level when dealing with fishing down has the effect of *under*estimating the fishing down effect (by 10–15%).[12] This gave us the first part of the answer to the puzzle posed by Caddy et al. We eventually came to see it as judo argument No. 1.

The second aspect of the taxonomic resolution problem, as seen by Caddy et al., was the degree of aggregation of the underlying FAO data, which they knew very well, given that one *alia* (R.J.R. Grainger) was the chief of FAO's Information, Data, and Statistics Unit.

Unfortunately, while they point out that of all fish caught, "about 20% cannot even be assigned to the level of Family," they fail to mention that there is a clear pattern to this reporting nonetheless: while high latitude countries (north and south) report about 90% of their catches to FAO at the level of species, low latitude countries (and China) tend to report them mostly as "mixed fishes" and other unhelpful categories (Pauly and Palomares 2005). The lack of taxonomic resolution occurs mainly in the tropics, and it has the effect of masking trends in trophic levels, as illustrated here from a high latitude area with well-disaggregated catches (Figure 2.8).

Figure 2.8 thus refutes the notion that, somehow, overaggregated data generate trends that can be misinterpreted as fishing down. The well-disaggregated data show the trend unequivocally. Thus the poor resolution from tropical countries most likely causes an *under*estimation of the strength of the fishing-down phenomenon. This was judo argument No. 2.

Interestingly, there is another form of overaggregation that causes similar problems: spatial overaggregation. This is detailed in Pauly and Palomares (2005), and illustrated in Figure 2.9 with data originating from FAO Area 31. These data, analyzed jointly, show mean trophic levels to be fluctuating with an upward trend; thus, one could conclude that fishing down does not occur in FAO Area 31 (Figure 2.9A). However, fishing down becomes very visible when FAO Area 31 is separated into two subareas: (a) from the Atlantic Mexico to Venezuela, an area whose catch (mainly of high-trophic-level fish) was strongly increasing in the 1990s (upper trend line in Figure 2.9B) and (b) the northern Gulf of Mexico, long exploited by US fleets of trawlers targeting (and reporting) shrimps and purse seiners targeting (and reporting) and Gulf menhaden, respectively, that is, low-trophic-level species (lower trend line on Figure 2.9B). Calculating a catch-weighted "mean" trophic level over two or more such starkly differing areas is not appropriate, as it results in aliasing (see also Box 1.2 and Pauly and Palomares 2005).[13]

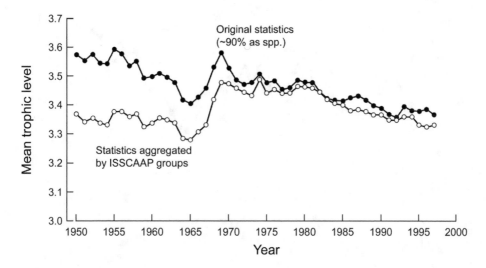

Figure 2.8. Illustrating the effect of taxonomic overaggregation on evidence for "fishing down." The black dots, based on highly disaggregated catch data (about 90% at species level) from the northeastern Atlantic (FAO Area 27; see also Figure 2.3C), document a clear decline of 0.04 per decade. The white dots represent the same data, now trendless, after aggregation into the large groups of the International Standard Statistical Classification of Aquatic Animals and Plants (ISSCAAP) used by FAO.

Landings Data as Ecosystem Indicators

Caddy et al. doubted that the relative abundance of species in landings reflected their relative abundance in the ecosystem. As we pointed out in our initial response, "Peruvian landings consist mainly of anchoveta because these are abundant in the Peruvian upwelling ecosystem, and Indonesian coastal fishers land ponyfishes because these are abundant on the Sunda Shelf. Off Newfoundland, Canada, where cod was targeted until it recently collapsed, a fishery for invertebrates has recently developed. It can be safely expected that Newfoundland's future landing statistics will reflect the species shift that occurred in the ecosystem around that island."

So we thought at the time. But was that really the case? Understanding this issue in specific cases is very much a matter of taking into account the capabilities of the fishing gears that are deployed. Again, as we originally responded, the "correspondence between relative abundance in the landing and in the ecosystems was not the rule before fisheries became globalized, and only selected species were exploited by near-shore gear. Now, with inshore, offshore- and distant-water fleets competing to supply increasingly integrated global markets, abundant species are exploited wherever they occur (Grainger and Garcia 1996), and landings will tend to reflect their relative abundance." Also, in many cases, markets were more selective, and they concentrated

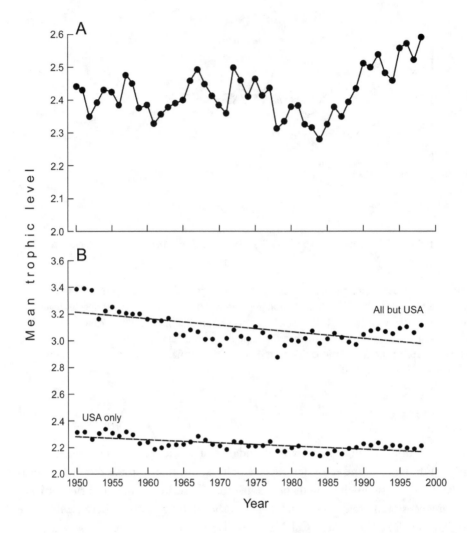

Figure 2.9. Illustrating the effect of spatial overaggregation on evidence for "fishing down." (A) Trendless time series of mean trophic levels, based on FAO landing data from the whole of the western central Atlantic (FAO Area 31); (B) the same data, after separation into two subsets: the northern Gulf of Mexico (USA), and the rest. This reveals two trend lines previously masked by aliasing (see text).

on a few, locally known and appreciated species, as opposed to the situation prevailing nowadays where, at least in wealthy countries, the markets display vast arrays of seafood of uncertain origin and identity (Jacquet and Pauly 2008), all the way down to the surimi "fish" sticks now available at my local supermarket in Sète.

As it happened, the correspondence hypothesis was tested, by comparing the species composition of commercial trawler catches with the species relative abundance in the Celtic Sea ecosystem, as assessed by trawl surveys (Pinnegar et al. 2002).

The results were surprising—or, upon reflection, perhaps not: the decline of mean trophic level was less pronounced in the commercial catch than in the survey catch (i.e., in the ecosystem). Pinnegar et al. (2002) attributed this to the skippers' attempts to maintain a high-value catch by targeting larger, high-trophic-level fishes (which are more valuable).

If this can be generalized—and there is no a priori reason why it cannot be—this means that a declining trend in the mean trophic level of landings would, other things being equal, actually *under*estimate the extent of fishing down that occurs in the ecosystem. This became judo argument No. 3.

Aquaculture Development

Caddy et al. (1998b) suggested that "aquaculture production may not have been fully excluded from [our] analysis" and suggested, along with Pinnegar et al. (2003), that fishing down may be an artifact resulting from the statistics we used. Caddy et al., indeed, pointed out that, at the time, FAO data did not distinguish well between fisheries catches and aquaculture production, particularly in the Mediterranean. (Again: they were more critical of FAO data than anyone before had dared to be!) Yet here, also, they failed to consider what the effect of such overaggregation would be on trends of mean trophic levels. In the case of mariculture, in countries other than China (see Chapter 3 for this special case), the trend has been, over the last 50 years, that the farming of low-trophic-level animals (oysters, mussels, etc.) has remained stagnant or has even declined, and the culture of high-trophic-level fishes such as salmon, seabass, and bluefin tuna (or shrimps fed fishmeal!) has grown rapidly. The result is that we are, with regard to aquaculture, "farming up the food web" (Pauly et al. 2001b), particularly in the Mediterranean (Stergiou et al. 2009).[14] Thus, if we had failed to exclude farmed organisms, this would have resulted in *under*estimating the fishing-down effect. This was judo argument No. 4.

Eutrophication and Other Bottom-Up Effects

In the Mediterranean, then the focus of J.F. Caddy's work, and in other coastal seas, river discharges in the second half of the 20th century included huge amounts of nutrients because of changing land use, especially the application of fertilizers (see, e.g., Caddy et al. 1995; Maranger et al. 2008). These nutrients, particularly in the oligotrophic Mediterranean, have increased primary production, which can be expected to have led to increases of the biomass of low-trophic-level small pelagic fish populations (anchovies, sardines, mackerels, etc.).

Thus, a situation can be conceived wherein the mean trophic level of the ensemble of fisheries exploiting such a modified ecosystem would decline because of the increased contribution of small pelagics to their catch, even though the contribution of larger, higher-trophic-level fish might not change. Such an "addition without depletion" scenario is, indeed, the basis of "fishing *through* the food web," proposed by Essington et al. (2006) as an extension of, or corrective to, fishing *down*.[15]

There are two ways to evaluate the validity of this proposition. One rather straightforward approach is to compute mean trophic levels from catch or landing data that exclude fish with low trophic levels (Pauly and Watson 2005), whose short-time fluctuations often obscure long-term trends.

Figure 2.10, based on Bhathal and Pauly (2008) displays time series of such truncated mean trophic levels for each of the States and Union Territories of India, and each shows a clear declining trend. On the other hand, a similar study, which disaggregated the catch data only between the west and east coasts of India, and which failed to exclude lower-trophic-level fish (and tuna, see above), showed evidence of fishing down only for the west coast of India (Vivekanandan et al. 2005). Figure 2.10 shows that fishing down, at least in this case, was not due to increasing catches of small pelagics (here Indian oil sardines) or other species with trophic levels < 3.25.

Similarly, Stergiou (2005), based on catch data from Greek waters, analyzed trophic level "slices." His results were the same: addition of low-trophic-level fish to the catch (as required by fishing "through" the food web) is not necessary for fishing down to occur, and the higher-trophic fish declined fastest. This was judo argument No. 5.

Another way to examine the possibility of a bottom-up effect is to account explicitly for the fact that marine ecosystems operate as pyramids, wherein the primary production generated at trophic level 1 moves up toward the higher trophic levels, with a huge fraction of that production being wasted for the maintenance, reproduction, and other activities of the animals in the ecosystems (Cury et al. 2003; see also the discussion of transfer efficiency in Chapter 1). Thus, notwithstanding a human preference for catching and consuming large predators, deliberately fishing down should enable more of an ecosystem's biological production to be captured by fishing. However, any decline in the mean trophic level of fisheries catches should be matched by an ecologically appropriate increase in these catches, the appropriateness of that increase being determined by the transfer efficiency between tropic levels. Thus, a Fishing-in-Balance (FiB) index can be defined that

- will remain constant (remains = 0) if trophic level changes are matched by "ecologically correct" changes in catch;
- will increase (>0) if either (a) "bottom-up effect" occurs, for example, increase in primary production in the Mediterranean (which triggered Caddy et al.'s concerns),

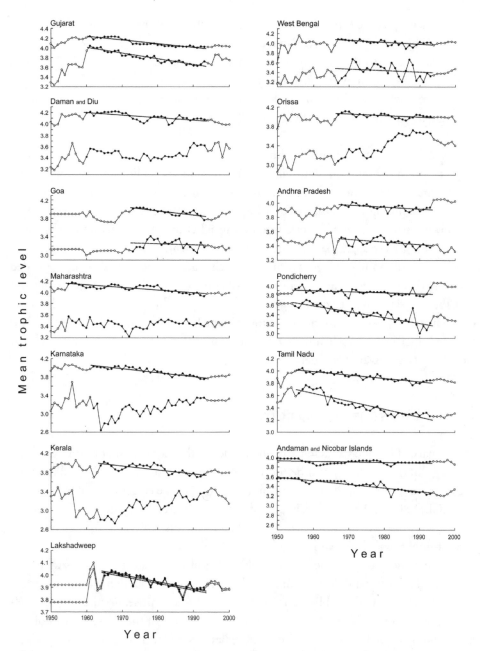

Figure 2.10. Trends in mean trophic level in Indian States and Union Territories (UT), with panels arranged such as to roughly reflect geography. In each panel and for each year, the upper dots represent mean trophic levels excluding organisms with trophic levels >3.25, while the lower dots include all fish and invertebrates. For all States and UT, the upper full dots provide strong evidence of "fishing down" in India and confirm that lower-trophic-level organisms are best deleted from analyses of fishing down. The open dots identify periods when the sampling of fisheries statistics was insufficient for analysis (see Bhathal and Pauly 2008 for details).

or (b) a geographic expansion of the fishery occurs, and the "ecosystem" that is exploited by the fishery has been in fact expanded;

- will decrease (<0) if discarding occurs that is not considered in the "catches," or if the fisheries withdraw so much biomass from the ecosystem that its functioning is impaired (which corresponds to the "backward-bending curves" mentioned in the original contribution on "fishing down").

Box 2.2 presents the equation defining the FiB index and, as well, the basis of its interpretation in spatial terms, and Figure 2.11 presents an application example. Other examples may be found in Sherman and Hempel (2008) and at www.seaaroundus .org. Jointly, they illustrate that accounting for bottom-up effect in the context of fishing down is straightforward and leads to new insights about the functioning of marine ecosystems. This was the sixth, and final, judo argument.

All in all, the suite of investigations in which Caddy et al. led us to engage ended up strengthening, and nuancing, our original conclusions. The fact of the matter was that the first signal we detected, suggesting a trophic decline, was muted by the noise of the many different confounding factors to which Caddy et al. had pointed us. It is to such criticisms, even when from detractors, as Caddy et al. indisputably were, that we scientists are invariably indebted.

The CBD and Its "Marine Trophic Index"

In February 2004, the Conference of the Parties to the Convention on Biological Diversity (CBD) identified a number of indicators to monitor progress toward reaching the target to "achieve by 2010 a significant reduction in the current rate of biodiversity loss" (CBD 2004). Among the eight biodiversity indicators suggested for "immediate testing" was the "marine trophic index" (MTI), which is how the CBD renamed the mean trophic level (Pauly and Watson 2005).

The metamorphosis of the mean trophic level of fisheries catches, a research tool whose declining trend indicated functional changes in the underlying ecosystem, to a biodiversity index endorsed by 188 countries came as a surprise, despite ongoing collaboration with the CBD (Vierros and Pauly 2004). One of the consequences of this metamorphosis, however, was more work: what is simpler, if you have to report on your country's progress on the MTI, than to send me an e-mail asking me what do?[16]

Of course, we were very proud of the fact that our findings had ultimately led to an important new policy initiative, one that promised to make fisheries management more protective of the health of marine ecosystems, which until then had not been a factor at all in reporting or management. For this we have to thank, not merely Caddy

Box 2.2

Definition and Interpretations of the Fishing-in-Balance (FiB) Index

The equation defining the Fishing-in-Balance (FiB) index is

$$FiB_k = \log [Y_k \cdot (1/TE)^{TL_k})] - \log [Y_0 \cdot (1/TE)^{TL_0}] \qquad (1)$$

where Y_k is the catch in year k; TL_k and TL_0 are the mean trophic level of the catch in year k and 0, respectively; TE is the mean transfer efficiency; and 0 refers to any year used as a baseline to normalize the index (Pauly et al. 2000). When TE is set at 0.1 (see Chapter 1), the FiB index simplifies to $FiB_k = \log [Y_k \cdot 10^{TL_k}] - \log [Y_0 \cdot 10^{TL_0}]$.

The FiB index, as defined above, increases if catches increase faster than would be predicted by TL declines, and it decreases if increasing catches fail to compensate for a decrease in TL. This is due to the fact that, in the absence of geographic expansion or contraction and with an ecosystem that has maintained its structural integrity, movement of the fisheries down the food web should result in increased catches (and conversely for increasing TL), with the FiB index remaining constant.

Examination of various case histories (Pauly et al. 2000; Pauly and Palomares 2005; Bhathal and Pauly 2008) shows that the FiB index increases where geographic expansion of the fisheries is known to have occurred. This can be made explicit by normalizing the FiB index for the areas covered by the fishery in a given year (A_k), relative to the area covered in the baseline year (A_0), which leads to an area-weighted FiB index, here called Balance-in-Fishing (BiF) index:

$$BiF_k = \log[Y_k \cdot (1/TE)TL_k \cdot A_0] - \log[Y_0 \cdot (1/TE) \cdot TL_0 \cdot A_k] \qquad (2)$$

Thus, we can define what might be called a spatial expansion factor, A_k/A_0:

$$A_k/A_0 = 10^{(FiB_k - BiF_k)} \qquad (3)$$

Given accurate catch data and correct estimates of TE, TL_r and A_k, the value of the BiF index should (by definition) remain zero through time. Thus, equation (3) can be simplified, for any year, to

$$\text{Expansion factor} = 10^{FiB} \qquad (4)$$

This holds, however, only on the assumption that this expansion involves areas with the same productivity as that of the area exploited in the baseline year (i.e., A_k can support, on a per-area basis, the same catch as A_0). This assumption, though likely never to be strictly met in practice, may not be too unrealistic when successive depth ranges of a smoothly sloping shelf are compared. In fact, if productivity declines with depth, this will cause the expansion to be underestimated. Thus, expansion factors estimated in this fashion would tend to be conservative.

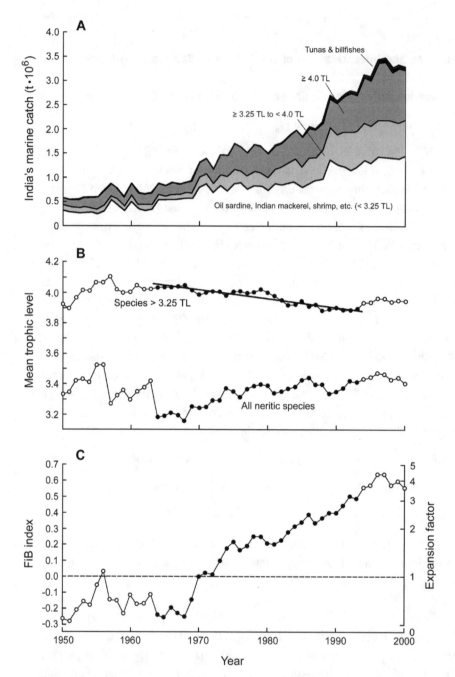

Figure 2.11. Basic trends in the Indian fisheries, suggesting that the increase of catches in (A) are due to a geographic expansion. The decline in trophic level (B) started in the early 1970s, when the FiB index started to increase beyond its values in the 1950s and 1960s, when the largely un-motorized fisheries stagnated. The FiB values reached in the 1990s suggest an approximately four-fold expansion relative to the area fished in 1970 (see Box 2.2 and Bhathal and Pauly 2008).

et al. and others of our scientific colleagues, but also the media, though our gratitude was tempered by no small amount of chagrin.

The Jellyfish Sandwich

The coverage of "Fishing Down Marine Food Webs" by newspaper and other periodicals was huge, often in articles with lurid titles such as "Global Collapse of Fisheries Feared" (*Vancouver Sun*) or "The Rape of the Sea" (*New Scientist*).

Two aspects of this coverage may be mentioned here. The first is that the *New York Times* of February 10, 1998, ran a thoughtful article entitled "Man Moves Down in the Marine Food Chain, Creating Havoc" in its Science[17] section (Stevens 1998). The article was accurate,[18] informative, and well written, and I understood then that there are professionals with communication skills that most scientists lack. It is then that I began responding to journalists' phone calls, and learned to communicate with them, so they could translate my work into something that could be read and appreciated by a wide range of audiences.[19]

Another aspect of the press coverage we received that may be worth mentioning is that in several interviews, I had suggested, half in jest, that if we were to continue fishing all the way down, we would "soon be eating plankton soup and jellyfish burgers." Somehow, the hyperbole caught people's attention, and a second wave of articles appeared on the prospect of jellyfish taking over the world, and becoming the only available seafood.

I had almost forgotten about this when, in 2000, I received an invitation to give the keynote address at the First International Conference on Jellyfish Blooms.[20] I did not accept then: what did I know, really, about gelatinous zooplankton? But I accepted an invitation to speak seven years later, at the Second International Conference on Jellyfish Blooms, in Brisbane, Australia, if only to find out why I was being invited.

The scientists who specialize in the biology and ecology of jellyfish are few, and traditionally they had access to the mass media only when a pile of dead jellyfish made the local beach unsightly. Jellyfish burgers, on the other hand, were the latest new thing, and they increased the public profile of jellyfish research; hence the invitations, I was told in Brisbane. But the spotlight in which they found themselves was soon outshone by a larger one, occasioned by another paper, from Jeremy Jackson.

In July 2001, his paper, coauthored by 16 colleagues, was featured on the cover of *Science* (a big deal!), and it showed the fishing-down effect to be far older than the less than 50 years we had considered. Also, it must be seen as part of a sequence of events with far more ominous effects than mere changes in the relative composition of fisheries catches (Jackson et al. 2001).

This seminal work, which documented human-induced local and global extinction of marine mammals (e.g., Steller's sea cows), seabirds (e.g., the great auk), and reptiles (local populations of all sea turtles) long before the onset of the industrial age, forcefully made the point that our major mode of interaction with the ocean (grab all we can, and eat it) will, if left to its own devices, leave us with empty oceans. Or rather, they will be full of organisms we do not like, such as various microbes and harmful algae (Weiss 2007), and jellyfish (Pauly et al. 2009). In fact, it will be as if the evolutionary sequence had been reversed (Pauly 1979b; Parsons 1996), with the more derived, larger animals being replaced by simpler, smaller ones, all the way down to bacteria.

An example of an ecosystem where this future is already unfolding is the northern Benguela ecosystem, off Namibia, in southwestern Africa, where 15 million tonnes

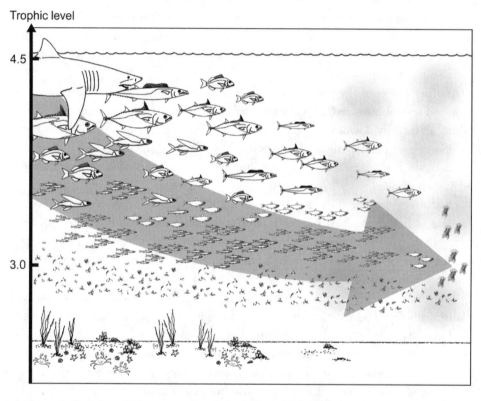

Figure 2.12. Schematic representation of the fishing down process. Fisheries usually start in the upper left corner, targeting large, long-lived fishes. Once they become scarce, the fisheries move to small fishes, often the prey and the juvenile of the larger fish, which decline even more. Eventually, the fisheries concentrate on small fish and invertebrates. At the same time, the use of trawls and other habitat-destroying gears tends to eliminate the filter- and deposit-feeding organisms on the sea floor. Together, these impacts lead to conditions conducive to jellyfish and harmful algae blooms. This is the "rise of slime" (see text).

of good fish, such as hake and sardine, have been replaced by 12 million tonnes of jellyfish (Lyman et al. 2006), which are now busy eating up the eggs and larvae of the remaining fishes. The frenetic fishing that occurred earlier in that ecosystem is bound to be linked to this change. And yes, the higher-trophic-level fish went down faster than the others, inducing fishing-down effects (Willemse and Pauly 2004a, 2004b).

Thus, I predict that we will see quite a few more versions of Figure 2.12, which is often used to illustrate "fishing down" in textbooks. Maybe it will be used to introduce a new name for this new era, the age of slime. I propose that it be called the "Myxocene."[21]

CHAPTER THREE

China and the World's Fisheries

Spring in Vancouver Island

The third article had its genesis on Vancouver Island, where in the spring of 2000 the newly founded *Sea Around Us* Project had its first review, by a group of accomplished fisheries scientists (Poon 2000). Richard Grainger was one of the reviewers; he was also one of the coauthors of the Caddy et al. critique of our paper on fishing down the food web (see Chapter 2) and, it may be mentioned again, chief of FAO's Information, Data, and Statistics Unit.[1]

Reg Watson and I where chatting with him when the subject of China's coastal fisheries came up. China's enormous catches produced a huge red anomaly on the first, tentative, global catch maps produced by the project.[2] Richard Grainger smiled knowingly: everybody in his group knew that China had been overreporting its marine catch for years, and FAO had household fish consumption surveys data confirming this. "But why doesn't FAO correct for this?" Again, he smiled knowingly: "You know how such things are…!" And later, he and Reg Watson agreed that Reg would get a small contract from FAO to estimate the extent of overreporting by China, and FAO would use their report to move China to correct their reporting. That was the plan; no one but the parties involved knew of this agreement, which is interesting in view of the positions taken by some of the actors. But I am again getting ahead of myself.

Given the ubiquity of the "fishing down" process presented in Chapter 2, and knowledge of declining catches from various areas throughout the world, it was, on the face of it, quite astonishing that the FAO should report that global catches through the 1990s were increasing. But it did,[3] and I smelled a rat.

It became clear that the major reason for the catch anomaly was the extremely rapid increase of reported landings from the People's Republic of China, which increased at a steadily accelerating pace, from about 4 million tonnes in the mid-1980s to over 15 million in 1999, when the Chinese government decreed a transition to a "zero growth policy," wherein catches would not be increased above their 1998 value.[4] Thus, given the confirmation of my doubts by Dr. Grainger, I used the opportunity of the Open Lecture for the International Council for the Exploration of the Sea (ICES), which I gave in Bruges in September of that same year. I highlighted a graph suggesting that the world catch was actually decreasing outside of Chinese waters, and I said that this was a good reason to think seriously about conservation (not that I got anywhere: this was a tough crowd). And upon my return, I pinned this graph above Reg's desk: this was going to be what we worked on.

The likely explanation for the anomalous catches in China was deliberate over-reporting, a feature common to the primary industries of command-and-control economies (Henderson et al. 2005). However, one cannot simply conclude anything from a propensity to malfeasance. One needs to demonstrate each specific case, as do criminologists, when they insist on the need for a potential culprit to have had (a) the motive, (b) the means, and (c) the opportunity to commit the deed (Doyle 1902; McCall-Smith 1999). In this case, this meant demonstrating (a) why anyone in China would benefit from overreporting fisheries catches, (b) that the statistical system in China allows for systematic overreporting to be perpetrated, and (c) that these opportunities are (or were) indeed taken. Moreover, to be credible, one would also have to quantify the extent of the overreporting.

Clearly, none of these demonstrations could be achieved without dedicated sleuth-ing.[5] An important part of this investigative work was conducted by Ms. Lei Pang, a Chinese national and graduate of the Monterey Institute of International Studies in California, who assembled from Chinese government Web sites, then translated, numerous official statements admitting systemic overreporting of (e.g., agricultural) production figures by Chinese officials—whose promotions are usually conditional on production increase in the sectors they supervise. At the largest, most inclusive scale, this includes reporting on the entire Chinese economy, where, for example, the overall economic growth rate is supposed to have been nearly twice that of electricity genera-tion, improbable as it is (Thurow 2007).[6]

The production of "watery statistics" (as the Chinese expression goes) that this engenders is a problem that the Chinese government had recognized and tried to stamp out since the very founding of the People's Republic of China in 1949 (Kwong 1997). Indeed, these largely unsuccessful efforts, and their dreadful consequences—including large-scale famines, reported in literary form by Chang (1991)[7]—have led to an extensive English language literature on this specific form of corruption, which also could be brought to bear on the issue of exaggerated fisheries catches (see Pang and Pauly 2001).

The other part of this work, including quantifying Chinese overreporting, was done by Reg Watson of the *Sea Around Us* Project, as part of his development of a method for generating the fine-grained maps of global fishing catches (Watson 2001).[8]

The work was then put together, written up, and sent to FAO in the summer of 2001, to be published in one of their Circulars or other series, where it would have remained accessible only to experts, that is, to people who already "knew" of this overreporting by China. Thinking this would be interesting to a wider audience, Reg Watson and I also summarized the report into a shorter contribution, which we submitted to *Nature*, where it had the good luck of passing editorial muster[9] and had reasonable anonymous reviewers, notably the respected British fisheries scientist J.G. Shepherd.[10]

Our contribution was published in November 2001, and all hell broke loose, not least because *Nature* had provided our contribution with an introduction by Pearson (2001) that also featured a cartoonish rendition of a nonexistent "FAO Fish Catch Committee." In the cartoon, four committee members watch with dismay as a fifth appears to be telling them a "tall fish story," with a big grin and his arms stretched out. Here is the text of a news piece by Pearson which preceded our contribution:

NATURE

China Caught Out as Model Shows Net Fall in Fish

H. Pearson | Vol. 467

Despite a wealth of local evidence suggesting that world fish stocks are in peril, the United Nations' Food and Agriculture Organization (FAO) has consistently reported that global catch sizes are stable or rising.

But new research suggests that the FAO statistics, which have encouraged investment in the fishing industry, may have been distorted by exaggerated estimates of catch sizes by some countries—particularly China. The reality, say the researchers, is that fish stocks have declined alarmingly over the past decade.

The FAO classifies more than 70% of major marine fisheries as fully or overexploited. Many populations, such as the North Atlantic cod, have already crashed. But despite these warning signs, FAO statistics for total global fish catches have increased since 1950. These seemingly anomalous figures were put down to the discovery of new stocks.

Reg Watson and Daniel Pauly of the University of British Columbia in Vancouver, Canada, have now re-analysed the FAO statistics, using information about factors such as food abundance and water depth to predict catch levels. They report the results in this issue of *Nature* [see below]. Their model mirrored the FAO figures in most regions, but China's reported catches—which account for around 15% of the global harvest—are twice the predicted figure. If Watson and Pauly are correct, and China has over-reported its catches, world fish stocks have actually declined by more than 10% since 1988.

"The results are stunning," says Jane Lubchenco, a marine biologist at Oregon State University in Corvallis.[11] "We're on a trajectory of significant decline." Only a drastic overhaul of fishery management can halt the global trend, she says.

Local Chinese officials, whose promotion is linked to their ability to exceed production targets in the country's socialist economy, may be responsible for the over-reporting, believes Pauly. The central Chinese government seemed to acknowledge the problem in 1998, when it placed a cap on the figures reported to the FAO.

Dropping stocks threaten the fishing industry and world food production. Fish provides more than 15% of the world's animal protein, and many developing countries in particular rely heavily on fish catches."

Hear is the text of our report:

Systematic Distortion in World Fisheries Catch Trends[12]
R. Watson and D. Pauly | Vol. 414

Over 75% of the world marine fisheries catch (over 80 million tonnes per year) is sold on international markets (FAO 2000a), in contrast to other food commodities (such as rice; Maclean 1997). At present, only one institution, the Food and Agriculture Organization of the United Nations (FAO) maintains global fisheries statistics. As an intergovernmental organization, however, FAO must generally rely on the statistics provided by member countries, even if it is doubtful that these correspond to reality. Here we show that misreporting by countries with large fisheries, combined with the large and widely fluctuating catch of species such as the Peruvian anchoveta, can cause globally spurious trends. Such trends influence unwise investment decisions by firms in the fishing sector and by banks, and prevent the effective management of international fisheries.

World fisheries catches have greatly increased since 1950, when the FAO of the United Nations began reporting global figures (FAO 2000b). The reported catch increases were greatest in the 1960s, when the traditional fishing grounds of the North Atlantic and North Pacific became fully exploited and new fisheries opened at lower latitudes and in the Southern Hemisphere. Global catches increased more slowly after the 1972 collapse of the Peruvian anchoveta fishery (Muck 1989), the first fishery collapse that had repercussions on global supply and prices of fishmeal (Figure 3.1A). Even taking into account the variability of the anchoveta, global catches were therefore widely expected to plateau in the 1990s at values of around 80 million tonnes, especially as this figure, combined with estimated discards of 16–40 million tonnes (Alverson et al. 1994), matched the global potential estimates

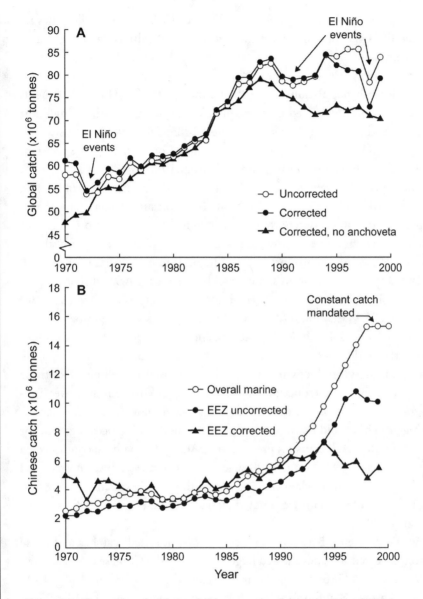

Figure 3.1. Time series of global and Chinese marine fisheries catches (1950 to present). (A) Global reported catch, with and without the highly variable Peruvian anchoveta. Uncorrected figures are from FAO (2000b); corrected values were obtained by replacing FAO figures by estimates from (B). The response to the 1982–83 El Niño/Southern Oscillation (ENSO) is not visible as anchoveta biomass levels, and hence catches were still very low from the effect of the previous ENSO in 1972 (Muck 1989). (B) Reported Chinese catches (from China's exclusive economic zone [EEZ] and distant water fisheries) increased exponentially from the mid-1980s to 1998, when the "zero growth policy" was introduced. The corrected values for the Chinese EEZ were estimated from the general linear model described in the Methods section.

NATURE

published since the 1960s (Pauly 1996) [see Chapter 1]. Yet the global catches reported by the FAO generally increased through the 1990s, driven largely by catch reports from China.

These reports appear suspicious for the following three reasons:

(1) The major fish populations along the Chinese coast for which assessments were available had been classified as overexploited decades ago, and fishing effort has since continued to climb (Huang and Walters 1983; Tang 1989); (2) Estimates of catch per unit of effort based on official catch and effort statistics were constant in the Yellow, East China and South China seas from 1980 to 1995 (Chen 1999), that is, during a period of continually increasing fishing effort and reported catches, and in contrast to declining abundance estimates based on survey data (Tang 1989); (3) Re-expressing the officially reported catches from Chinese waters on a per-area basis leads to catches far higher than would be expected by comparison with similar areas (in terms of latitude, depth, primary production) in other parts of the world. We investigated the third reason in some detail by generating world fisheries catch maps on the basis of FAO fisheries catch statistics for every year since 1950 (see Figure 3.2A for a 1998 example). A statistical model was used to describe relationships between oceanographic and other factors, and the mapped catch. Most high-catch areas of the world were correctly predicted by the model. These areas typically had very high primary productivity rates driven by coastal upwellings, like those off Peru, supporting a large reduction fishery for the planktivorous anchoveta *Engraulis ringens* (Muck 1989). The exception was the waters along the Chinese coast. Here, the high catches could not be explained by the model using oceanographic or other factors. Yet the catch statistics provided to FAO by China have continued to increase from the mid-1980s until 1998 when, under domestic and international criticism, the government proclaimed a "zero-growth policy" explicitly stating that reported catches would remain frozen at their 1998 value (Figure 3.1B) (see also Pang and Pauly 2001).

Mapping the difference between expected (that is, modeled) catches and those mapped from reported statistics showed large areas along the Chinese coast that had differences greater than 5 tonnes km^{-2} $year^{-1}$. Overall, the statistical model for 1999 predicted a catch of 5.5 million tonnes, against an official report of 10.1 million tonnes (see Figure 3.1B for earlier years). Although China was not the only FAO member country reporting relatively high catches, their large absolute value strongly affects the global total.

For a number of obvious reasons, fishers usually tend to under-report their catches, and consequently, most countries can be presumed to under-report their catches to FAO. Thus we wondered why China should differ from most other countries in this way. We believe that explanation lies in China's socialist economy, in

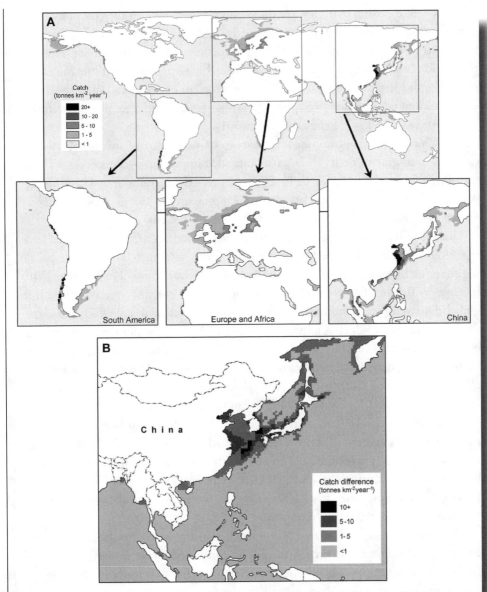

Figure 3.2. Maps used to correct Chinese marine fisheries catch in Figure 3.1(B). (A) Map of global catches reported by FAO for 1998, generated by the rule-based algorithm described in the Methods section. We note the anomalously high values along the Chinese coast, comparable in intensity to the extremely productive Peruvian upwelling system. (B) Map of differences in southeast and northeast Asia between the catches reported in (A) and those predicted by the model described in the Methods section. [The three inserts, featuring areas of details, were not in the original version of this graph, which, however, was in color.]

which the state entities that monitor the economy are also given the task of increasing its output (Kwong 1997). Until recently, Chinese officials, at all levels, have tended to be promoted on the basis of production increases from their areas or production units (Kwong 1997). This practice, which originated with the founding of the People's Republic of China in 1949, became more widespread since the onset of agricultural reforms that freed the agricultural sector from state directives in the late 1970s (Kwong 1997; Pang and Pauly 2001).

The Chinese central government appears to be well aware of this problem, and its 1998 "zero-growth policy" was partly intended to prevent over-reporting. Thus, the official fisheries catches for 1999–2000 are precisely the same as in 1998 (Figure 3.1B), and will be for the next few years. Such measures, although well motivated, do not inspire confidence in official statistics, past or present.

The substitution of the more realistic estimated series of Chinese catches into the FAO fisheries statistics led to global catch estimates which, although fluctuating, have tended to decline by 0.36 million tonnes year^{-1} since 1988 (rather than increase by 0.33 million tonnes year^{-1}, as suggested by the uncorrected data). The global downward trend becomes clearer when the catches of a single species, the Peruvian anchoveta, which is known to be affected by El Niño/Southern Oscillation events, is subtracted (see Figure 3.1A). In this case, a significant ($P<0.01$), and so far undocumented downward trend of 0.66 million tonnes year^{-1} becomes apparent for all other species and fisheries. This is consistent with other accounts of worldwide declines of fisheries (Botsford et al. 1997; Pauly et al. 1998a).

Ironically, it is likely that, at the lowest levels (individual fishers), catches are under-reported in China as elsewhere in the world. The production targets caused these reports to be exaggerated. At some times these two distortions may perhaps have cancelled each other out, and an accurate report of catches may have been submitted to FAO. Since the early 1990s, however, the exaggerations have apparently far exceeded any initial under-reporting.

The greatest impact of inflated global catch statistics is the complacency that it engenders. There seems little need for public concern, or intervention by international agencies, if the world's fisheries are keeping pace with people's needs. If, however, as the adjusted figures demonstrate, the catches of world fisheries are in general decline, then there is a clear need to act. The oceans should continue to provide for a substantial portion of the world's protein needs. The present trends of overfishing, wide-scale disruption of coastal habitats and the rapid expansion of non-sustainable aquaculture enterprises, however, threaten the world's food security.

Methods

Data processing involved a disaggregation of global fisheries catch statistics firstly into detailed taxonomic groups, and then into fine-scale spatial cells (a half-degree of latitude by a half-degree of longitude), using a variety of databases and systematic rules (Watson et al. 2001a). The spatially disaggregated catches provided the basis for a general linear model of fisheries catches (see below). The model predicted the likely catches in the spatial cells in the Chinese exclusive economic zone (EEZ), thus providing an estimate of Chinese catches (including Hong Kong and Macau, but excluding Taiwan).

Data Sources

Fisheries catch statistics were provided by the FAO (FAO 2000b) and as "Atlas of Tuna and Billfish Catches" (FAO 2000c). The spatial cells were described by depth (US National Geophysical Data Center), primary productivity (JRC 2000), biogeochemical provinces (Longhurst 1998b), the presence of ice (US National Snow and Ice Data Center, http://www.nsidc.org), surface temperature from an online marine atlas (NOAA 2000) and an upwelling index calculated [by Villy Christensen] for each cell by multiplying negative deviations in surface temperature (from the average for that latitude and ocean) by the primary productivity in that cell.[a] Fishing access rights were determined using maps of the exclusive economic zones (EEZ) of coastal states (Veridian Information Solutions 2000) and a database of fishing access agreements (FAO 1999).

Taxonomic Disaggregation

The fisheries statistics of several nations commonly include a large fraction of catches in "miscellaneous" categories. Chinese catches so reported were disaggregated on the basis of the breakdown provided by its two nearest maritime neighbours with detailed marine fisheries statistics (Taiwan and South Korea; Watson et al. 2001a). Assigning catches to lower taxa allowed the use of biological information in the spatial disaggregation process.

a. An upwelling index was needed because the catches from upwelling areas, notably the "big four" (the Peru, California, Canaries, and Benguela upwellings; Bakun 1990), are much higher than would be predicted from the other variables considered here (depth, distance from the coast, etc.). Our upwelling index was derived from temperature anomalies, i.e., we assumed that upwellings occur where the mean sea surface temperature is markedly lower than the ocean-specific latitudinal mean.

Spatial Disaggregation

A database of the global distribution of commercial fisheries species was developed using information from a variety of sources including the FAO, FishBase (Froese and Pauly 2000) and experts on various resource species or groups. Some distributions were specific; others provided depth or latitudinal limits, or simple presence/absence data. The spatial disaggregation process determined the intersection set of spatial cells within the broad statistical area for which the statistics were provided to FAO, the global distribution of the reported species, and the cells to which the reporting nation had access through fishing agreements (Watson et al. 2001a). The reported catch tonnage was then proportioned within this set of cells (Figure 3.2A).

Catch Predictions

A general linear model was developed in the software package S-Plus20 (Insightful Corporation 2001). The model relates log fisheries catch (in tonnes km^{-2} $year^{-1}$) for each cell (the dependent variable) to depth, primary productivity, ice cover, surface temperature, latitude, distance from shore, upwelling index (the continuous predictor variables), 33 oceanic biogeochemical provinces and one global coastal "biome" including most of the area covered by the world's EEZs, including China's (the categorical predictor variables). Fishing effort was not used in the prediction and catches were assumed to be generally close to their maximum biologically sustainable limits. The additive and variance stabilizing transformation (AVAS) routine of S-Plus20 was used to identify transformations ensuring linearity between the dependent and explanatory variables, and the model was then used to predict the catch from each spatial cell. Those from Chinese waters were combined, then compared with the catches obtained from the rule-based spatial disaggregation described above (Figure 3.2B).

Trend Analyses

The estimates of recent trends of global catch were estimated by linear regression of catch versus year, for the period from 1988 (highest catches, anchoveta excluded) to 1999 (last year with FAO data), for uncorrected global marine catches, global marine catches adjusted for Chinese over-reporting, and adjusted catches minus the catch of Peruvian anchoveta.

The Economist, the FAO, and the World

The impact of our contribution in the mass media was huge, and is outlined below. But first, we deal with *The Economist,* a British magazine that is widely read, including in the USA where nearly half of its readership resides, and where it presents itself as a bridge to Europe and the rest of the world. One of its affectations is that its articles have no byline, that is, one is not supposed to know who authored them. Besides this, *The Economist* leans editorially toward a more laissez-faire approach to government, toward outsourcing jobs to the cheapest labor markets and other forms of globalization. It generally finds fault with environmental regulations (too expensive) and limits on what multinational corporations can do, and international agreements: this should all be left to market forces.[13] Thus, *The Economist'*s reporting on the United Nations and its specialized agencies—including the FAO—generally has a negative slant. It was therefore not surprising that *The Economist'*s coverage of our contribution (Loder 2001) bashed FAO. Loder's account is presented first, before FAO's own response, because the vehemence of that response is partly due, I think, to their being the target of both *Nature'*s editorial cartoon and the article in *The Economist.* The latter now follows:

THE ECONOMIST

The World's Fish Catch May Be Much Smaller than Previously Thought

N. Loder | December 2001

Fishery statistics tend, just like fish, to be rather slippery. Lying by individuals, industries and countries is expected by the body that has collated global fishery statistics for the past half-century. But the Food and Agriculture Organisation[14] had assumed that, unless everybody lied at the same time and in the same direction, discrepancies in the global figures would pretty much cancel each other out.

Unfortunately, this approach overlooked the possibility that a single large contributor might be lying spectacularly. And according to Reg Watson and Daniel Pauly, two researchers at the University of British Columbia in Vancouver, that is what China has been doing for at least ten years. This, they say, has masked a big downward trend in the global fish catch.

Their research was prompted by the observation that local fisheries around the world were collapsing, yet the global catch, which was expected to plateau in the 1990s at around 80m tonnes per year, was slowly increasing. Taking the FAO's fish-catch statistics since the 1950s, the researchers worked out the relationship between catch and various oceanographic and environmental factors, such as depth of the ocean, latitude, ice cover, surface temperature and distance from the shore.

After verifying that their model was able to predict the location of most high-catch regions of the world, they went on to create a global map of the difference between expected (or modelled) catches, and officially reported statistics. This revealed a shocking discrepancy. In China, a catch of 5.5m tonnes was expected in 1999; but the official figure was 10.1m tonnes.

When the pair replaced official statistics with estimates, the global catch showed a wobbly downward trend, shrinking by some 360,000 tonnes every year since 1988. And when they removed the catches of a single, highly fluctuating species, the Peruvian anchoveta, the data revealed a strong and consistent downturn, of 660,000 tonnes a year. In other words, contrary to official figures suggesting that the marine catch has been slowly growing for the past few years, it has in fact been in decline.

That the Chinese figures are unreliable is hardly surprising, since until recently Chinese officials were promoted on the basis of production increases. What is surprising is that such a distortion of global statistics might be possible. The FAO offers several defences. One is that these new findings, published in this week's *Nature*, are based on modelling, which does not prove anything. The suggestion that China might be cooking the books is not new. The FAO says it has been suspicious of the Chinese figures for the past six years.

Richard Grainger, the FAO's chief statistician, argues that global figures are not important, because fisheries are managed at a regional level. This means that any inaccuracies in the Chinese figures would affect only China and not perceptions of the state of other world fisheries. Because China is not a great importer or exporter of fish, the food-security implications are limited to the region. Anyway, he says, few people look at global figures without reference to regional trends.

Not So Many Fish in the Sea

Andy Rosenberg, a fisheries scientist at the University of New Hampshire, disagrees. He says that graphs showing a stable global catch are often shown at international meetings, not least by the FAO. Indeed, on the first page of the FAO's most recent annual report, the global fish catch is described as remaining "relatively stable."

Dr Rosenberg also says that many countries assume that, as long as the overall picture remains healthy, fisheries management is a problem for the long term. As long as global volumes are rising or stable, it seems reasonable to conclude that the exhaustion of local fishing grounds has been balanced by the opening of new grounds farther afield. The new research suggests that this is wrong.

If the global catch is declining, despite the unprecedented effort being made to maintain production, stocks must be in decline too. What can be done? Some look to fish farming, or aquaculture, as a way of maintaining production. In the short term, this may work. But most farmed fish are fed a diet consisting mainly of fish taken out of the ocean. So although aquaculture may boost edible fish production, it is ultimately limited by marine fish resources.

One novel approach attempts to bring

consumer pressure to bear. Unilever, the world's largest buyer of frozen seafood, set up the Marine Stewardship Council (MSC), in conjunction with the World Wildlife Fund, in 1998. The MSC sets environmental standards for sustainable and well-managed fisheries and awards a quality mark to those that make the grade. Unilever says it will buy all its fish from sustainable fisheries by 2005.

Many other options have been proposed to deal with the problem of overfishing, such as reducing the capacity of fishing fleets, setting up marine reserves, removing government subsidies or assigning property rights to individuals or groups of fishermen to provide an incentive for good stock-management practices. The problem with this latter approach is that it requires elaborate and expensive policing. And if stocks are stable, as the FAO's figures suggest, why bother?

As with global warming, governments will take action only when the urgency of the situation has become fully apparent. By pointing out that a stable supply of marine fish can no longer be taken for granted, Dr. Watson and Dr. Pauly have raised an important alarm.

FAO's response was twofold: First, the report we had submitted to them, and which was slowly moving toward publication, was "pulled," and ceased to exist. Thus, it was published as *Fisheries Centre Research Report* (Watson et al. 2001a).

Second, FAO provided a rather lengthy reply, not so much to the substance of the article itself (whose technical contents were, for good reasons, never contested), but mainly to the comments published by *The Economist*. Here, the main points of FAO's long reply, as posted on their Web site (FAO 2002), are presented, with deletions indicated by "[…]" and with author-and-year-style references replacing links to the online versions of the articles cited. Here we go:

A recent scientific article in *Nature* (Watson and Pauly 2001) indicated that China's marine capture fishery production for 1995–1999 has been overstated in Chinese statistics submitted to and published by FAO. Th[is contribution] states that a consequence of this is that global marine capture fishery production—excluding Peruvian anchoveta—has probably been declining since 1988 rather than remaining fairly constant as indicated by the statistics. According to the[se] authors […] this would have led to understating the degradation of world fisheries and wrong policy and investment decisions. The issue has been subsequently taken up in a number of newspapers and web media including *The Economist* (Loder 2001). While usefully drawing attention of the wider public on the importance of reliable statistics for fisheries management and monitoring, the articles also reflected a number of misconceptions.

The FAO went on listing three areas about which they provided a detailed response: (a) their understanding of Chinese statistics; (b) FAO's role in gathering, interpreting, and disseminating global fisheries statistics; and (c) their perception of the likely impact of Chinese overreporting on the management of global fisheries and food security issues. Significant quotes pertaining to these three points follow:

V1. FAO's Understanding

It is not the first time that scientists report the "finding" that China's fishery statistics overestimate production.[15] In fact, several Chinese scientists have previously referred to the problem. FAO has been concerned about China's agriculture and fisheries statistics for several years and has been working with China to rectify these. Following the first national agricultural census in China, undertaken with the collaboration of FAO in 1997, statistics for meat production were revised downward by about 25%. [...]

About six years ago, apparent discrepancies between per caput food fish supply data derived from fish production and trade statistics on the one hand and consumption figures derived from household surveys on the other started to grow. FAO drew this issue to the attention of the responsible Chinese authority, the Bureau of Fisheries (BOF) of the Ministry of Agriculture. Since then several meetings, missions and three national seminars (e.g., FAO 2001b) have been jointly organized. [...] FAO has initiated a number of follow-up activities in conjunction with several Chinese institutions, including the establishment of a pilot sample survey data collection scheme for one county in 2002. Further, an assessment will be made of the quantity of unprocessed fish landings which are used for direct feed in aquaculture. The quantities are believed to be very large and, at present, are wrongly assumed to go for human consumption. The problem is therefore known and action is being taken on it.

The important message here was that FAO did not dispute the results. This is emphasized here because later things get murkier.

Next is the FAO's role in global fisheries statistics. This must be read carefully because this role is crucial. Indeed, in many of the world's developing countries, the fisheries statistics submitted to FAO by officials in one or the other ministry (External Trade, or Finance, or Agriculture, etc.) are often the only set of national statistics local fisheries scientists will ever see.[16] What is strangest about it is the way FAO, here, "owns up" to the statistics' imperfection, although their staff will, in private, admit that they do not have the personnel to do half of what they claim to do (which is already a lot):

FAO

2. FAO's Role in Fishery Statistics

Most press news related to the UBC article stated wrongly that FAO can only accept data from the countries without any possibility of checking or improving them. While this would conveniently limit the liability of FAO about any error in the data it publishes, it does not reflect reality. [...] FAO interacts with countries to explore problems and try to resolve them, though the process can be sensitive and slow. When countries are not responsive to FAO inquiries, FAO estimates are applied unilaterally. Occasionally, when countries offer little supportive explanation of suspect statistics, reported statistics are set aside and FAO estimates published. This action too can be sensitive and provocative, but does sometimes spur corrective action.[17] As pointed out elsewhere in this article, China is working with FAO to try to address these issues.[18]

[...] Global reviews of the state of stocks elaborated and published by FAO are not based on catch statistics as primary source of information because there are often more direct indicators of the state of resources than the catch. Primary information used is obtained directly from the working groups of the FAO and non-FAO regional fishery organizations (RFOs) and other formal arrangements, scientific literature (scientific journals, theses, etc.), supplemented by information from industry magazines, and fishery-independent information such as trade data. Where RFOs do not exist, such as in the Northwest Pacific, there are bilateral assessment processes (e.g., among China, Japan and Republic of Korea) which can be built upon. Where data do not exist, on discards for instance, estimates are made on a one-off basis (e.g., Alverson et al. 1994), using consultant experts or through dedicated expert consultations.[19] In the areas in which FAO has not yet had the means to work effectively, e.g., production from illegal fishing, there is no information at all at global level. However, all such data are only available for certain areas or certain years. The great advantage of FAO's catch statistics is that they are global in coverage, have complete time series since 1950 and are regularly updated, and so they are used to provide overview trends in fisheries by region (e.g., FAO 1994) and to provide resource status indicators when other data are lacking (e.g., FAO 1997a).

During the last decade, financial support for the development and maintenance of national fishery statistical systems has decreased sharply in real terms, while statistical requirements

FAO

have been increasing dramatically for by-catch and discards, fishing capacity, illegal fishing, vessels authorized to fish in the high seas, economic data (costs, revenues, prices, subsidies), employment, management systems, inventories of stocks and fisheries, aquaculture, etc.

Despite FAO's efforts, the fishery data available are not fully reliable. The outcome is far from perfect in terms of coverage, timeliness, and quality. Data are submitted to FAO often with one or two years of delay. The proportion of catch identified at the level of individual species has tended to decrease with time, and the percentage of "unidentified fish" in the declarations has increased as fisheries diversified and large stocks were depleted. Stock assessment working groups provide a good means for screening catch data, but the frequency of stock assessment in many developing regions has dropped for want of human and financial resources. The general availability of data has not really improved during the last two decades. Statistics from artisanal and subsistence fisheries are still a concern and many key statistics are missing, e.g., economic and social data, discards, fishing capacity. [...] The FAO Fisheries Department believes that working with the countries is the only way to improve fishery statistics, primarily to meet national needs with regard to food security and fisheries management, but also those of regional fishery bodies and FAO. Without reliable statistics, effective fisheries management and policy-making are impossible, with serious negative implications at the national and regional levels. Unfortunately, the rehabilitation of major national data collection schemes to provide reliable statistics is necessarily a slow process.[20]

Here, the statement of note is in the penultimate sentence: "Without reliable statistics, effective fisheries management and policy-making are impossible, with serious negative implications at the national and regional levels." Which is true. And this is emphasized in the following:

FAO

Global statistics have been used in FAO's Committee on Fisheries (COFI) and other global fora. Despite the shortcomings with the data, they have accompanied national efforts for institutional change in FAO, in the Commission for Sustainable Development (CSD) and in the UN General Assembly. The UN Fish Stocks Agreement, the FAO Compliance Agreement, the FAO Code of Conduct[21] and its four International Plans of Action are the result of such action at global level. All of this was based on the recognition of the fisheries problems at the highest levels of government and industry and certainly do not reflect "*complacency*" on the part of FAO.

Based on this and other activities, FAO also notes that:

FAO

member countries brought into force the 1982 Convention on the Law of the Sea (1994) and the UN Fish Stocks Agreement (2001); adopted the FAO Compliance Agreement (1993) and the FAO Code of Conduct for Responsible Fisheries (1995); aligned their national laws with these instruments; adopted and started implementing the precautionary approach; accepted the concept of ecosystem-based fisheries management;[22] adopted four international plans of action (on management of fishing capacity, shark fisheries management, reducing by-catch of seabirds in longline fisheries, illegal/unreported/unregulated fishing) and will consider the possibility of adopting one on improvement of information on status and trends of capture fisheries.

However, hidden in this blizzard of words (and this is much shortened) is the notion that, actually, accurate catch statistics may not be that important:

FAO

3. Implications for Management, Policy and Food Security

[...] It has been argued that the potential error in statistics may have affected the "vision" of the state of stocks. FAO stresses that in order to elaborate its judgment on the global state of stocks, as published for instance in FAO (1997a), it uses mainly the results of direct assessments complemented by analysis of catch data. Such assessments are collected from regional fishery body working groups, national centres of excellence, scientific publications, grey literature reports, etc., supplemented by ancillary information.

Following a complete overhaul of the FAO fishery production database, two series of analyses were made (Grainger and Garcia 1996; Garcia and Newton 1997). The first one (Grainger and Garcia 1996) was exploratory and surprisingly showed a remarkable coherence between the data and the well-known history of fisheries and yielded, *inter alia*, a dramatic and unique picture of the progression of overfishing since 1950 [...]. The second led to the development by FAO of a unique bio-economic model of world fisheries (Garcia and Newton 1997) which, for the first time, demonstrated: (a) the global overfishing since the 1980s of the most valuable species; (b) the fact that by the mid-1990s the world potential was reached; and (c) that a large amount of subsidies were probably used to maintain the world fishing fleet which operated with an overcapacity of 30–50%. This should demonstrate that, despite any potential error in Chinese production figures, the overall negative trend in global stocks and the poor economic

FAO

performance of world fisheries has been made sufficiently apparent, described and brought to the attention of FAO members as well as the public at large. As a matter of fact, these conclusions have been very widely quoted by scientists, reproduced in the media and used by all leading NGOs.

Regarding more specifically the Northwest Pacific Region, possible errors in Chinese statistics would only affect the assessment of stocks exploited by China and its neighbours. FAO studied the changes in trend in this area, as well as in the other fishing areas, also from the point of view of the production per shelf area (Caddy et al. 1998a) and trophic level (Caddy and Garibaldi 2000). Caddy et al. (1998a) documented regional declining trends in production and provided some explanations for them. FAO also published a report by a Chinese scientist (Chen 1999), which evidenced that many marine resources of China have become seriously over-exploited.

I must interrupt this, and do it in the main text of this book, as the last sentence in the above quote is too important to be clarified only for people who also read endnotes: the paper published by Chen (1999) as *FAO Fisheries Circular* not only shows that Chinese coastal fish populations are "seriously over-exploited" but proudly presents prima facie evidence for the large-scale manufacture of catch data, in the form of three figures exhibiting 15 years of almost linearly increasing fishing effort, increasing catches, but miraculously constant catch per unit of effort (see Figure 4 in Pang and Pauly 2001). Did anyone at FAO notice? Maybe not; maybe this just slipped through. The terrifying question is, "If yes, what does it imply?" One dares not think this thing through. But we must go on:

FAO

In the *Nature* article, the authors not only corrected the world catch for the Chinese "error" but also eliminated from it the large anchoveta catch, the reality of which is otherwise not questioned, allegedly because its known fluctuations would "hide" the decline.[23] This arbitrarily aggravates the decrease in production possibly resulting from a Chinese bias in the world database and the resulting "decrease" is presented as a novelty. In reality the problem posed by large stocks of fluctuating pelagic and semi-pelagic species had already been raised by FAO in 1994 (Garcia and Newton 1997). Subtracting from the total production the catch of the five main oscillating species (anchoveta, Chilean horse mackerel, Japanese and American pilchards and Alaska pollock), most of which are utilized for reduction to feeds, FAO demonstrated that the large, high-value species landings decreased *since the early 1980s* and that all these species were globally over-fished by that time with an overcapacity of 30%. The "news" offered in *Nature* therefore simply confirms an important information already available for many years from FAO.[24]

The ultimate section of FAO's response, that is, on the role of statistics and their analysis in terms of policy and food security, is extremely long and cannot be reproduced here. Moreover, it doesn't respond to the contents of our contribution, but instead deals largely with *The Economist*'s accusation of "complacency," notably by listing all the things that FAO does to improve the status of global fisheries. Thus, I shall limit myself here to citing their main points. Here we go:

FAO

Contrary to allegations in some media reports, FAO has not been "complacent." Despite difficulties with the data, the Organization has correctly stated that food fish production, despite its apparent stability, had not been keeping pace with population growth and has warned that fish protein supply per caput had already declined in some areas. [...] (Figure 3.3).

It may be disconcerting that, despite the growing environmental impacts of aquaculture and overfishing, and taking account of all voluntary or involuntary errors in statistics, the overall contribution of fisheries and aquaculture to food security does not appear significantly threatened in the medium term. The two messages on the state of resources and the state of food security, should, however, not be confused. The poor state of resources should be of serious concern today and should indeed have been corrected during the last two decades. The overall contribution of fisheries and aquaculture to food supplies is not yet a concern. The future availability to the poorest stratum of the world population is a concern because of the increase of external trade and prices. The future quality of seafood is of concern because of growing pollution and inappropriate culture techniques. All these concerns, for the moment and for the near future, are not global and are relevant only at disaggregated (national or sub-national) level, for some production systems, in some areas, and for some strata of the population. FAO has, for instance, warned in SOFIA 2000 (FAO 2000a) that the fish supply per caput in Africa has been declining and will decline even further in the future if African countries cannot better manage their resources and/or increase aquaculture production.[25]

There is no doubt, however, that a global decline in marine catches would have sent an even stronger message to the countries, the public and NGOs

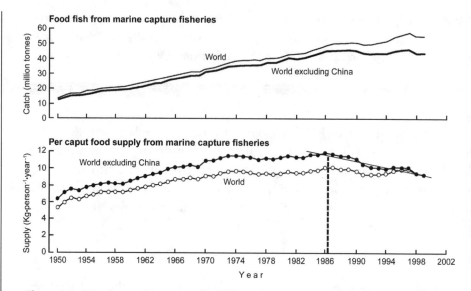

Figure 3.3. "Absolute and per caput food fish production from marine capture for the world and the world excluding China (1950–1999)" as presented on the FAO Web site (see FAO 2002).

than the one already repeatedly sent by the Organization since the early 1990s. The amount, direction, and specifics of the Chinese bias, if any, need to be ascertained. In the meantime, as already done in SOFIA 2000, most relevant FAO analyses will be presented as a precautionary measure separately for China and the rest of the world.[26]

Chinese Responses

There were numerous responses by China—which is not surprising, as the country is huge. However, what is interesting is that in the first responses of startled officials to the Beijing-based foreign journalists who called them, there was a concession that there was a problem with catch data reporting (along with insistence that it had been fixed in the meantime). Erik Eckholm, of the *New York Times*, one of the journalists who called early, put this delicately: "Chinese officials have acknowledged problems with past data and with the bureaucratic incentives" (Eckholm 2001). Later, the official position hardened, and these concessions were never made again. Also, we received some nasty e-mails,[27] which tended to elaborate on the official position, as reported by the Xinhua News Agency (also known as the "New China News Agency"), the news agency of the People's Republic of China.

The article below, translated by Ms. L. Pang, was authored by Du Shuangqi, Xinhua News Agency, December 18, 2001, and had the title "*Nature* Report Unfounded: China Did Not Over-Report Fisheries Catch Statistics." The text follows:

Yang Jian, director of the Bureau of Fisheries of the Department of Agriculture, claims today, in an interview, that Chinese official fisheries catch statistics are basically accurate and that China did not supply inflated catch numbers to the international community.

Yang Jian states: China's work on fisheries catch statistics has strictly adhered to [China's] Statistical Law and followed the statistical procedures approved by the National Bureau of Statistics; Chinese statistical approach may differ from the sampling survey methods widely used in other countries; admittedly, our statistical process is not perfect; the Chinese government is more than willing to collaborate with international organizations to make improvements in this area.

Two Canadian scientists Reg Watson and Daniel Pauly published an article in British journal *Nature* on November 29 this year, saying that the reason why global fisheries catch statistics published by the Food and Agriculture Organization of the United Nations are continuing to increase, despite steady depletion of fisheries resources worldwide, is that some countries, including China, have over-reported fisheries catch statistics to the organization since 1980, thereby misleading the international community about the state of fisheries resources.

Yang Jian says that the increase in Chinese catches in the past twenty years is reasonable. With aquatic products being one of the first commodities released from governmental price control, market demand has substantially driven industry development. The total number of Chinese fishing boats increased from 49,000 in 1980 to 280,000 last year and fisheries remains a profitable sector of agriculture today.

"Additionally, Chinese fisheries have their own characteristics. Owing to our distinct food culture and diet tradition and due to our socio-economic conditions, aquatic products that are not utilized by many developed countries, such as certain crustaceans and jelly fish, account for a significant proportion of our catch," he said. "The statistical model used by the Canadian scientists may not accommodate these characteristics."

In spite of this, the Chinese government has consistently put great emphasis on protecting fishery resources. Presently, China enforces annual comprehensive fishing moratoriums with durations of two to three months. This year, over 130,000 large fishing boats that may inflict damage to fishery resources entered moratoriums, involving over one million fishermen.

Yang Jian states: last year Chinese catch dropped 1.35 per cent from the previous year. Larger decreases in catch are possible for this year. Thanks to tight control over newly built fishing boats by the fishery departments, there has been basically no increase in the number of boats since 1998. Yet China plans a further reduction of 30,000 fishing boats in the next five years.

Unofficial accounts were more nuanced. Thus, the following account is a translation of an article by Li Hujun, published in the *Southern Weekend*, on December 13, 2001. Ms. L. Pang, who translated the article, also suggests that "the *Southern Weekend* is a very popular (for its bold, liberal contents) weekly newspaper based in Guangzhou, Guangdong Province. Their reporters and managers often have troubles with the government." Here it is:

SOUTHERN WEEKEND

China Over-Report Fisheries Catch Statistics?

Li Hujun
(Ms. Lei Pang, trans.) | December 13, 2001

Nature recently published an article saying China has submitted fraudulent and unreasonably high fisheries catch statistics to the Food and Agriculture Organization of the United Nations. This reporter visited the relevant authorities and experts in an attempt to seek out the truth.

On November 29, in [an article published by] the internationally acclaimed academic journal *Nature*, two Canadian researchers claimed that China has committed fraud in submitting fisheries catch statistics to the Food and Agriculture Organization of the United Nations. The claim has brought stormy reactions from everywhere.

For some years, the state of global fisheries resources has been dire. Yet, quite perplexingly, the global catch statistics published by FAO [have] been in a steady rise. Watson and Pauly provided an explanation for this [puzzle], in the British journal *Nature*. The so-called steady increase is but a false impression and the clue lies in the fraudulent statistics provided by China.

According to the two researchers, the major fishing grounds along the Chinese coast have been over-fished since as early as the 1980s and related [past resource] surveys have also shown a declining trend of Chinese fish abundance; compared with other areas in the world, Chinese catch per unit area is surprisingly high. The researchers' suspicion did not end here. They set out to build a statistical model using ocean environmental factors to assess the catch levels of national waters over the past years. Reported statistics of most countries correspond with the results predicted by the model. [The statistics from] China [are] the only exception. Between mid-1980s and 1998, China's reported statistics show a steady increase of catch from about 4 million tonnes to 15 million tonnes, twice as high as the result predicted by the model.

Watson and Pauly believe that China is not the only country that has mis-reported fisheries catch. Out of tax consideration, fishermen tend to under-report catches to government; hence, statistics submitted by some countries may be lower than the actual levels. What is strange is that China is the only country that has provided unreasonably high catches statistics. They suspect this has to do with China's unusual economic management system: catch increases, or fishery growth, is often viewed as an official accomplishment. Consequently some officials tend to over-report production levels.

China occupies a prominent position in world fisheries, with its catch accounting for fifteen per cent of the global harvest. Precisely because of this, *Nature* states, if the Canadian researchers are correct, global fisheries catches, rather than steadily rising, have been declining ten per cent a year since more than ten years ago. Marine biologist Jane Lubchenco says, the [research] result is "stunning," showing that global fishery catch has long been on a trajectory of decline and the problem of resource depletion has long been pressing.

Did China Over-Report Fisheries Catch Statistics After All?

A relevant individual from the Bureau of Fisheries in the Department of Agriculture has claimed that China's statistics are recognized by the Food and Agriculture Organization of the United Nations and that the article in *Nature* represents nothing more than a conjecture made by some individual experts and is but one opinion. He told this reporter that certain experts merely infer from a statistical model without any knowledge of the concrete situation in China. Thus, following declining abundance in large and small yellow croakers, China developed new fisheries for stocks that were never previously fished.

This individual believes that a linkage between catch increases and the promotion of officials may exist, but not in the Bureau of Fisheries. Furthermore, [he pointed out,] China put forward a "zero growth" policy specifying that catch levels must not exceed the 1998 level. Also, China has been enforcing annual three-month fishing moratoriums in an effort to protect fishery resources.

But the explanations provided by the Bureau of Fisheries do not appear to fully answer the questions raised by Watson and Pauly, because these two researchers distrust the local officials who submitted the original data, not the officials from the Bureau of Fisheries. In fact, the "zero growth" policy is also considered by Watson and Pauly who view it as evidence of the central government of China acknowledging the problem of overly high catch statistics, although this [zero growth] policy cannot guarantee accuracy of official statistics.

What is the real Chinese catch? According to sources, the voice of suspicion has been heard not only abroad, domestic experts have also expressed doubts about the credibility of fisheries catch statistics. A relevant individual in the Bureau of Fisheries admits that discrepancies do exist in Chinese catches. The sizes and locations of the discrepancies are still being studied. The official also pointed out that China's statistical system differs greatly from those in foreign countries. Whereas the norm abroad is sampling surveys followed by mathematical models to obtain final statistical results, in China the norm is [distribution from the] top down of comprehensive forms to be used for reporting by the lowest administrative level. The Bureau of Fisheries is currently exploring ways to improve the designs of its index system and specific methodologies for fisheries catches statistics. Next year, in some trial locations, sampling surveys will be implemented, along with other procedures similar to those abroad.

Professor Jin Yongjin of the Department of Statistics at Renmin University of China told this reporter that "comprehensive form" reporting represents a holdover from the socialist planning economy, in which, theoretically, all original data are truthful and the final statistical results can be assumed more accurate than those generated by sampling surveys. However, when a link is made between statistical figures and an official's administrative achievement, the phenomenon wherein "statistics produce officials and officials produce

statistics" will tend to occur. Therefore, in a market economy, the use of comprehensive form reporting is generally replaced by the use of sampling surveys, which provide various ways to check on the reliability of the original data. Mature sampling techniques are also more effective in controlling errors.

[Watson and Pauly] emphasize that if Chinese fishery authorities are really lying, and if steady growth of global fisheries catches is really a false impression, adverse consequences will certainly follow. For instance, the public, led to believe that there is still plentiful fish in our oceans, will be blindly optimistic; those in fishing and finance industries may be misled to make bad investment decisions; and international fishery policy making may also be affected.

So the debate continues on the reliability of the statistics provided to FAO by China. "Fraudulent statistics are more harmful than insufficient statistics, because they can be misleading," said Jin Yongjin.

Taiwan, since 1949, has been an ideological counterpoint to the People's Republic of China, and hence it is not surprising that Taiwan-based newspapers, here the *Liberty Times*, should have used a story of statistical malfeasance in China to boost the Taiwanese view of things (Yan 2001). Ms. L. Pang, who translated this article, informed us that its author, Yan Hong Yang, is "a Taiwanese who is no friend of China and who is an associate professor of marine fishery biology (sensory biology and behavior) in the US. He wrote an article which contains what he called 'excerpts' from your contribution. However, his is a mix of problematic paraphrasing of your text, mixed into his anti-China rhetoric. The way he put words in your mouth is both stunning and amusing. [... However], if he had not inserted his own interpretations and were not so radically anti-China, he would have done quite a nice job toward the public understanding of science. I was struck by his ability of making it easy for ordinary people to understand the technical aspects of your contribution. After all, this won him two teaching awards at US universities in the past decade."

So here is the article:

LIBERTY TIMES

China, a Nation Good at Statistical Falsification

Yan Hong Yang

(Ms. Lei Pang, trans.) | December 8, 2001

On November 29, 2001, the 414th volume of *Nature*, the most authoritative journal of natural sciences, published an academic contribution titled "Intentionally distorted world fisheries catches" (in English "Systematic distortions in world fisheries catch trends"), coauthored by Reg Watson and Daniel Pauly, professors of the University of British Columbia in Canada. For those in Taiwan who are dazed for having contracted the "China fever," this contribution is well worth being used as a dose of antipyretic. This author is hereby providing excerpts for reference.

The Food and Agriculture Organization of the United Nations is the single international organization responsible for global fisheries statistics. It relies on all member states to supply it with fisheries data. The credibility of these fisheries catch statistics is entirely dependent on whether the reporting countries have reported truthfully. The two professors considered the fisheries statistics submitted by China as rather "suspicious" and they lay out three points of doubt to question their credibility. First, based on the fisheries surveys conducted by

China's own fisheries biologists along the Chinese coast in the past twenty years, all of the fishing grounds in the area have been "over-fished." As a result, fishermen have had to fish with increased effort [for an equal amount of catch], as can be inferred from the ever-rising fishing effort.

Second, Chinese official statistics for catches of the Yellow Sea, the East China Sea, and the South China Sea show rather constant annual catch per unit of effort during the fifteen years between 1980 and 1995, but surveys conducted in the same period by fishery biology researchers show an annual decline of fisheries resources in the above-mentioned areas. This means that a marked difference exists between two groups of statistics—the official statistics and academic statistics.

Third, calculation based on the official catches statistics published by the Chinese government indicates catches on a per-area basis that are far higher than other areas of the world, which is even more suspicious. To further investigate this last aspect, the two academics designed a mathematical model, using FAO world fisheries data since 1950, to predict relationships between catches and oceanographic factors (including depth, primary productivity, biogeochemical factors, distribution of sea ice, sea surface temperature, and upwelling). Based on model analyses, they discovered that the model could correctly predict all high-catch areas around the world. These high-catch areas almost all exist in waters associated with upwellings, where high primary productivity, thanks to the availability of abundant nutrients, is able to sustain high yield fisheries. The large, upwelling-induced anchovies fishing ground occurring along the Peruvian coast is one such example. What their mathematical model cannot explain is that there is no correlation whatsoever between the catches statistics published by China and the [environmental] factors characterizing the fishing ground.

In addition, comparing the fisheries statistics of China with the statistics of the Southeast and Northeast Asian fishing grounds, the annual catch per square kilometer in Chinese waters can be five tonnes higher than that observed in other Asian waters, which, based on fishery biology, is entirely impossible. Furthermore, based on model predictions, Chinese catch for the year 1999 could only be 5.5 million tonnes at most, but official statistics show a catch as high as 11.1 million tonnes. Put in another way, the official catch statistics are twice as high as the possible catch. The authors believed that the mentality in which statistics were falsified is directly associated with the ideology of the "socialist economy." To please their superiors, those in charge [of economic activities or of statistics] created a false picture of "a very good situation" by way of "injecting water" [into the statistics, to bloat them]. The authors further observed that it would not have mattered had China only wished to cheat on its own people behind closed doors. The problem is that, based on their estimates, global fisheries catches have been declining by 0.36 million tonnes per year in average, but thanks to the "water-injected" Chinese over-reported statistics, global fisheries catches now appear to have increased 0.33 million tonnes per year instead. Fraudulent catches statistics as such have two negative impacts: first, no conservation will be undertaken for fishing grounds and fisheries resources, due to the misperception of plentiful fisheries resources; second, misled by the statistics that have gone through "water injection," industries, to pursue profit, would invest huge sums of money to build more fishing vessels. The consequence of this sort of vicious circle would be accelerated depletion of fisheries resources [...].

Right at the time when the Chinese or Shanghai "investment fever" is getting fierce in Taiwan, this contribution in *Nature* is a blow to the head of, and a shout to the impetuous. I suggest that all those who desire to get rich in China should first examine the Chinese economic development data at hand and figure out that a good part of what composes "a vast panorama of very good scenery" has gone through meticulous "water injection" by the Chinese authorities. Should you foolishly plunge head first into China, you could end up losing everything.

The Media, or How Everyone Likes a Different Sauce

As the preceding exchanges illustrate, the relationship between science and policy (and politics!) can be very tense. In the case of the FAO, politics clearly trumped science, not out of any malice on the part of the scientists, of course. Most fisheries scientists—and probably most other scientists working on natural resource exploitation—work in government agencies and are expected to remain silent, even in the face of obvious conflicts of interest.[28]

This gives a special responsibility to those scientists in universities, where—at least in some countries—the tenure system (in principle) allows them to speak about their findings without censorship or retaliation. In fact, they often speak, for this reason, on behalf of their colleagues who cannot. It is therefore bizarre that we should accept a scientific ethos, pushed by some of my colleagues, wherein all that we are supposed to do is respond to the short-term questions of fisheries managers (e.g., what should the total allowable catch be, for next year?), rather than (also) exploring alternatives to the status quo (e.g., what kind of fisheries could we have that would result in greater benefits to society?).

This may be one of the reasons, along with the "capture" by the fishing industry of the very agencies that are supposed to regulate them (well documented for Canada; see Rose 2008), why fisheries resources are in the parlous state described in Chapter 2.

As we learn more and more about the nature of the dynamic biological systems we study as scientists, and as the integrity of those systems becomes increasingly compromised, we have a new role to play. Jane Lubchenco, in a 2004 speech delivered to the US National Academy of Sciences, called for a new contract between science and society, one in which scientists play a more visible role in bringing their work to public attention and ultimately into policy. Our paper was in part an effort to do just that, as we became more and more aware that our findings had powerful policy implications.

Moreover, the public's need to know, which is crucial in democracies, also extends to "scientific" matters (not least because they pay, through their taxes, for much of the research in question). And who is better positioned than scientists to make sense of what they do?

However, scientists are often not good at thinking like journalists and finding the angle that can make their work interesting and accessible to large audiences. This is why universities have public information, or similar, offices. Or, if a story really rocks, the scientist behind it can team up with a journalist, at least for a time, to write something together, for example a book (see Cury and Miserey 2008).[29]

In my case, one of the first journalists I worked with was the energetic and ever enthusiastic Nancy Baron, who described the launching of the *Sea Around Us* Project for the *Vancouver Sun* (Baron 1999a) and designed the first issue of our newsletter (Baron 1999b). At the time, she was working as a freelance science journalist, writing for magazines and regularly for the *Vancouver Sun*. She also wrote a feature story on me[30] and the work of members of UBC's Fisheries Centre on marine reserves (Baron 1998), which was a finalist for a National Magazine Award and which drew other journalists' attention to both of us, locally and across Canada.

Nancy later moved to the USA and worked with various projects devoted to linking scientists to science journalists and, as of this writing, serves as outreach director for the Communication Partnership for Science and the Sea (COMPASS). She was with COMPASS in the fall of 2001 and, after examining our draft, decided to pull out the stops to see our contribution brought to the attention of a wide audience, despite its nerdy title: "Systematic Distortion...." (Who cares about "systematic distortions" of anything or even knows what that means?) She worked with us to tell the story of what we had revealed and why it was important. Then, she drafted a press release replete with quotes by Reg Watson and me, and by scientists not involved with the study (which journalists love, as it saves them the time to get their own quotes, and it points them to other experts on the subject—something they would not typically know), and distributed it to the many science journalists she knew, and many she didn't. In fact, she contacted each of them personally and explained the significance of our contribution, that is, the answer to "So what?" It required background and knowing some of the "story behind the story." Indeed, it was the context of this study that made it interesting.

Our contribution was published in November 2001, and in February 2002, Nancy could report the following results: "Over ninety-five articles and radio pieces were generated as a result of the media relations effort. This was especially remarkable given the shrinking space for science journalism following the September 11th tragedy. Placements included *The Economist*, the Associated Press, the *New York Times* (twice), the *International Herald Tribune*, the *San Francisco Chronicle*, *US News and World Report*, *New Scientist*, as well as radio pieces on National Public Radio, BBC, CBC and the Australian Broadcasting Corporation. International placements included China, the United Kingdom, Canada, Australia, France and Germany."

Those who love sausages usually do not want to know how sausages are made, and similarly, some readers may be shocked to read how press coverage is generated. But to stay with the sausage analogy: the best media relations cannot decide which sauce and side dishes people will choose to have with their sausage; you can at best suggest (or pitch) a sauce. In fact, now moving from gastronomy to ideology: people interpret what they read through the lenses of their previous beliefs, and this is why the response to our contribution differed so much between, say, *The Economist* and Taiwan's *Liberty Times*.

Also, the media are not monolithic. Each has its own audience to serve and to answer, for them, the "So what?" question, that is, why does it matter to my audience? One aspect of this story that made it so appealing was that it overturned the then conventional wisdom and common knowledge about the state of global fisheries (i.e., that catches were increasing, whatever they all said). To others, like the *New York Times*, it was the fact that overreporting rather than underreporting was at issue, which went against the norm. To others it was further evidence to add to a clarifying picture of global fisheries' being in trouble—supporting previous recent studies that showed other ways of knowing this (e.g., Jackson et al. 2001).

The lesson: in individually pitching to these media, it helps to know the angle each favors; a quick Google search can usually reveal the type of story each tends to report. Still, journalists and their organizations vary greatly, and one ultimately has to do the best that one can, then relinquish control and hope for the best.

CHAPTER FOUR

Sustainability

What Is Sustainability, Anyway?

A vast number of definitions exist for the words *sustainable* and *sustainability*, but to make any sense, these definitions must imply that the process or state supposed to be sustainable (i.e., to exhibit sustainability) could continue forever, or at least for a very long time. This is why the concept of "sustainable growth" is nonsensical (Frazier 1997), however noble the intention that led to it being coined (WCED 1987). On the other hand, the much-criticized maximum sustainable yield (MSY; Larkin 1977) at least had the advantage of being theoretically coherent (important for scientists!), even if it is intractable in practice.

Similarly, fisheries are supposed to be managed for sustainability. The previous chapters, however, show that they are not. Fisheries economists tell us this is because we do not have "rights-based" fishing (Hanneson 2005), or fishing based on "catch shares" (Costello et al. 2008), the penultimate and last euphemism, respectively, for the privatization of fisheries (Macinko and Bromley 2002; Pauly 2008). The fishing industry says that it is because fisheries managers do not listen to them (Kurlansky 2008), fisheries biologists say the same thing (Ludwig et al. 1993), and conservationists want more marine protected areas (as I do, too). What is not obvious, and a point developed in the following contribution, is that humans, collectively, have never acted in an explicitly sustainable fashion toward any type of natural resources. Indeed, what in the historic record appear to have been periods of "sustainability" were times where lack of technology, of capital, or of markets prevented the fishery from growing, and from destroying its resource base. We still need to invent sustainability, which is also the reason why our contribution had "towards sustainability" in its title.

This contribution was easy to do in that it was solicited, in this case by the editor of *Nature*, who wrote me, presumably, because of the contributions presented in the previous three chapters.[1] His idea was that fisheries should be covered, along with other food-producing enterprises, in a special section devoted to the "future of food" (Gee 2002).

Given that the members of the *Sea Around Us* Project cover a wide range of expertise, it was an obvious decision to involve them in the resulting review, as well as Tony Pitcher, who had written on the need to rebuild fisheries resources rather than "sustain" them in their present depleted state (Pitcher 2001), and Carl Walters, who had earlier coauthored an important paper highlighting the general non-sustainability of fisheries (Ludwig et al. 1993). It was thus ironic that one reviewer, misunderstanding the thrust of our contribution, tried to make it say that only "unregulated" fisheries lack sustainability. Yet one of the seemingly best-regulated fisheries in the world, that of northern cod in eastern Canada, for which immense amounts of scientific information was available and fed to sophisticated models, collapsed spectacularly in the early 1990s (Walters and Maguire 1996).

Accommodating this and other reviewers' comments was relatively easy, though it required a letter of 19 pages documenting changes, or the reasons for not changing what had been originally submitted.

Reviews do not present new results, and hence there was no media response, although I was asked to summarize and simplify it for outlets with different audiences (Pauly 2004a, 2006a). On the other hand, the response from the scientific community was all one could hope for, that is, our contribution is well cited (see Epilogue). Here it goes.

NATURE

Towards Sustainability in Global Fisheries[2]
D. Pauly et al. | Vol. 418

Fisheries have rarely been "sustainable." Rather, fishing has induced serial depletions, long masked by improved technology, geographic expansion and exploitation of previously spurned species lower in the food web. With global catches declining since the late 1980s, continuation of present trends will lead to supply shortfall, for which aquaculture cannot be expected to compensate, and may well exacerbate. Reducing fishing capacity to appropriate levels will require strong reductions of subsidies. Zoning the oceans into unfished marine reserves and areas with limited levels of fishing effort would allow sustainable fisheries, based on resources embedded in functional, diverse ecosystems.

Fishing is the catching of aquatic wildlife, the equivalent of hunting bison, deer and rabbits on land. Thus, it is not surprising that industrial-scale fishing should generally not be sustainable: industrial-scale hunting, on land, would not be, either. What is surprising rather, is how entrenched the notion is that unspecified "environmental change" caused, and continues to cause, the collapse of exploited fish populations. Examining the history of fishing and fisheries makes it abundantly clear that humans have had for thousands of years a major impact on target species and their supporting ecosystems (Jackson et al. 2001). Indeed, the archaeological literature contains many examples of ancient human fishing associated with gradual shifts, through time, to smaller sizes and the serial depletion of species that we now recognize as the symptoms of overfishing (Orensanz et al. 1998; Jackson et al. 2001). This literature supports the claim that, historically, fisheries have tended to be non-sustainable, although not unexpectedly there is a debate about the cause for this (Ludwig et al. 1993), and the exceptions (Rosenberg et al. 1993). The few uncontested historical examples of sustainable fisheries seem to occur where a superabundance of fish supported small human populations in challenging climates (Boyd 1990). Sustainability occurred where fish populations were naturally protected by having a large part of their distribution outside of the range of fishing operations. Hence, many large old fecund females, which contribute overwhelmingly to the egg production that renews fish populations, remained untouched. How important such females can be is illustrated by the example of a single ripe female red snapper, *Lutjanus campechanus*, of 61 cm and 12.5 kg, which contains the same number of eggs (9,300,000) as 212 females of 42 cm and 1.1 kg each (Bohnsack 1990). Where such natural protection was absent, that is, where the entire population was accessible to fishing gears, depletion ensued, even if the gear used seems inefficient in retrospect (Yellen et al. 1995; Pitcher 2001). This was usually masked, however, by the availability of other species to target, leading to early instances of depletions observable in the changing size and species composition of fish remains, for example, in middens (Wing 2001).

The fishing process became industrialized in the late nineteenth century when English fishers started operating steam trawlers,[3] soon rendered more effective by power winches and, after the First World War, diesel engines (Cushing 1987). The aftermath of the Second World War added another "peace dividend" to the industrialization of fishing: freezer trawlers, radar and acoustic fish finders. The fleets of the Northern Hemisphere were ready to take on the world.

Fisheries science advanced over this time as well: the two world wars had shown that strongly exploited fish populations, such as those of the North Sea, would recover most, if not all, of their previous abundance when released from fishing (Hardy 1956). This allowed the construction of models of single-species fish populations whose size is affected only by fishing pressure, expressed either as a fishing mortality rate (F, or catch/biomass ratio), or by a measure of fishing effort (f, for example, trawling hours per year) related to F through a catchability coefficient (q): $F = qf$ (Schaefer 1954; Beverton and Holt 1957). Here, q represents the fraction of a population caught by one unit of effort, directly expressing the effectiveness of a gear. Thus, q should be monitored as closely as fishing effort itself, if the impact of fishing on a given stock, as expressed by F, is to be evaluated. Technology changes tend to increase q, leading to increases referred to as "technology coefficient" (Garcia and Newton 1997), which quickly renders meaningless any attempts to limit fishing mortality by limiting only fishing effort.

The conclusion of these models, still in use even now (although in greatly modified forms; Box 4.1), is that adjusting fishing effort to some optimum level should generate "maximum sustainable" yield, a notion that the fishing industry and the regulatory agencies eagerly adopted—if only in theory (Mace 2001). In practice, optimum effort levels were very rarely implemented (the Pacific halibut fishery is one exception; Clark et al. 1999). Rather the fisheries expanded their reach, both offshore, by fishing deeper waters and remote sea mounts (Koslow et al. 1999), and by moving onto the then untapped resources of West Africa (Kaczynski and Fluharty 2002), Southeast Asia (Silvestre and Pauly 1997a), and other low-latitude and Southern Hemispheric regions (Thorpe and Bennett 2001).

Box 4.1

Single-Species Stock Assessments [*Box originally drafted by C.J. Walters*]

Single-species assessments have been performed since the early 1950s, when the founders of modern fisheries science (Schaefer 1954; Beverton and Holt 1957) attempted to equate the concept of sustainability with the notion of optimum fishing mortality, leading to some form of maximum sustainable yield. Most of these models, now much evolved from their original versions (some to baroque complexity, involving hundreds of free parameters), require catch-at-age data. Hence government laboratories, at least in developed countries, spend a large part of their budget on the routine acquisition and interpretation of catch and age-composition data. Yet, single-species assessment models and the related policies have not served us particularly well, due to at least four broad problems. First, assessment results, although implying limitation on levels of fishing mortality which would have helped maintain stocks if implemented, have often been ignored, on the excuse that they were not "precise enough" to use as evidence for economically painful restriction of fishing (the "burden of proof" problem; Perry et al. 1999). Second, the assessment methods have failed badly in a few important cases involving rapid stock declines, and in particular have led us to grossly underestimate the severity of the decline and the increasing ("depensatory") impacts of fishing during the decline (Walters and Maguire 1996).

Third, there has been insufficient attention in some cases to regulatory tactics: the assessments and models have provided reasonable overall targets for management (estimates of long-term sustainable harvest), but we have failed to implement and even develop effective short-term regulatory systems for achieving those targets (Perry et al. 1999).

Fourth, we have seen apparently severe violation of the assumptions usually made about "compensatory responses" in recruitment to reduction in spawning population size. We have usually assumed that decreasing egg production will result in improving juvenile survival (compensation) so that recruitment (eggs × survival) will not fall off rapidly during a stock decline and will hence tend to stop the decline. Some stocks have shown recruitment failure after severe decline, possibly associated with changes in feeding interactions that are becoming known as "cultivation/depensation" effects (Walters and Kitchell 2001). According to this phenomenon, adult predatory fish (such as cod) can control the abundance of potential predators and competitors of their juvenile offspring, but this control [is] lost when these predatory fish become scarce. This may well lead to alternate stable states of ecosystems, which has severe implications for fisheries management (Scheffer et al. 2001).

Jointly, these four broad problems imply a need to complement our single-species assessments by elements drawn from ecology, that is, to move towards ecosystem-based management. What this will consist of is not clearly established, although it is likely that, while retaining single-species models at its core, it will have to explicitly include trophic interaction between species (Walters et al. 2000), habitat impacts of various gears (Hall 1998), and a theory for dealing with the optimum placement and size of marine reserves (see main text). Ecosystem-based management will have to rely on the principles of, and lessons learnt from, single-species stock assessments, especially regarding the need to limit fishing mortality. It will certainly not be applicable in areas where effort or catch limits derived from single-species approaches cannot be implemented in the first place.

Fisheries Go Global

In 1950, the newly founded Food and Agriculture Organization (FAO) of the United Nations began collection of global statistics. Fisheries in the early 1950s were at the onset of a period of extremely rapid growth, both in the Northern

Hemisphere and along the coast of the countries of what is now known as the developing world. Everywhere that industrial-scale fishing (mainly trawling, but also purse seining and long-lining) was introduced, it competed with small-scale, or artisanal fisheries. This is especially true for tropical shallow waters (10–100 m), where artisanal fisheries targeting food fish for local consumption, and trawlers targeting shrimps for export, and discarding the associated by-catch, compete for the same resource (Pauly 1997).

Throughout the 1950s and 1960s, this huge increase of global fishing effort led to an increase in catches (Figure 4.1) so rapid that their trend exceeded human population growth, encouraging an entire generation of managers and politicians to believe that launching more boats would automatically lead to higher catches. The first collapse with global repercussions was that of the Peruvian anchoveta in 1971–1972, which is often perceived as having been caused by an El Niño event. However, much of the available evidence, including actual catches (about 18 million tonnes; Castillo and Mendo 1987) exceeding officially reported catches (12 million tonnes), suggest that overfishing was implicated as well.[4] But attributing the collapse of the Peruvian anchoveta to "environmental effects" allowed business as usual to continue and, in the mid-1970s, this led to the beginning of a decline in total catches from the North

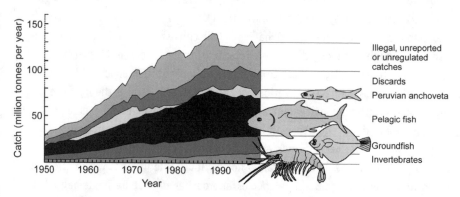

Figure 4.1. Estimated global fish landings 1950–1999. Figures for invertebrates, groundfish, pelagic fish, and Peruvian anchoveta are from FAO catch statistics, with adjustment for over-reporting from China (Watson and Pauly 2001). Fish caught but then discarded were not included in the FAO landings; data related to the early 1990s (Alverson et al. 1994) were made proportional to the FAO landings for other periods. Other illegal, unreported, or unregulated (IUU) catches (Bray 2000) were estimated by identifying, for each 5-year block, the dominant jurisdiction and gear use (and hence incentive for IUU; Forrest et al. 2001); reported catches were then raised by the percentage of IUU in major fisheries for each 5-year block. The resulting estimates of IUU are very tentative (note dotted y-axis), and we consider that complementing landings statistics with more reliable estimates of discards and IUU is crucial for a transition to ecosystem-based management.[5]

NATURE

Atlantic. The declining trend accelerated in the late 1980s and early 1990s when most of the cod stocks off New England and eastern Canada collapsed, ending fishing traditions reaching back for centuries (Myers et al. 1997).

Despite these collapses, the global expansion of effort continued (Garcia and Newton 1997) and trade in fish products intensified to the extent that they have now become some of the most globalized commodities, whose price increased much faster than the cost of living index (Sumaila 1999). In 1996, FAO published a chronicle of global fisheries showing that a rapidly increasing fraction of world catches originate from stocks that are depleted or collapsed, that is, "senescent" in FAO's parlance (Grainger and Garcia 1996). Yet, global catches seemed to continue, increasing through the 1990s according to official catch statistics. This surprising result was explained recently when massive over-reporting of marine fisheries catches by one single country, the People's Republic of China, was uncovered (Watson and Pauly 2001). Correcting for this showed that reported world fisheries landings have in fact been declining slowly since the late 1980s, by about 0.7 million tonnes per year.

Fisheries Impact on Ecosystem and Biodiversity

The position within ecosystems of the fishes and invertebrates landed by fisheries can be expressed by their trophic levels, expressing the number of steps they are removed from the algae (occupying a trophic level of 1) that fuel marine food webs (Box 4.2). Most food fishes have trophic levels ranging from 3.0 to 4.5, that is, from sardines feeding on zooplankton to large cod or tuna feeding on miscellaneous fishes. Thus, the observed global decline of 0.05–0.10 trophic levels per decade in global fisheries landings (Figure 4.2) is extremely worrisome, as it implies the gradual removal of large, long-lived fishes from the ecosystems of the world oceans. This is perhaps most clearly illustrated by a recent study in the North Atlantic showing that the biomass of predatory fishes (with a trophic level of 3.75 or more) declined by two-thirds through the second half of the twentieth century, even though this area was already severely depleted before the start of this time period (Christensen et al. 2003).

It may be argued that so-called "fishing down marine food webs" is both a good and an unavoidable thing, given a growing demand for fish (Pauly et al. 1998a). Indeed, the initial ecosystem reaction to the process may be a release from predation, where cascading effects may lead to increased catches (Daskalov 2002). Such effects are, however, seldom observed in marine ecosystems (Pace et al. 1999; Pinnegar et al. 2000), mainly because they do not function simply as a number of unconnected

Box 4.2

Trophic Levels as Indicators of Fisheries Impacts

There are many ways ecosystems can be described, for example in terms of the information that is exchanged as their components interact, or in terms of size spectra.[6] But perhaps the most straightforward way to describe ecosystems is in terms of the feeding interactions among their component species, which can be done by studying their stomach contents. A vast historical database of such published studies exists (Froese and Pauly 2006), which has enabled a number of useful generalizations to be made for ecosystem-based management of fisheries. One of these is that marine systems have herbivores (zooplankton) that are usually much smaller than the first-order carnivores (small fishes), which are themselves consumed by much larger piscivorous fishes, and so on. This is a significant difference from terrestrial systems, where, for example, wolves are smaller than the moose they prey on. Another generalization is that the organisms we have so far extracted from marine food webs have tended to play therein roles very different from those played by the terrestrial animals we consume. This can be shown in terms of their "trophic level" (TL), defined as 1 + the mean TL of their prey.

Thus, in marine systems we have: algae at the bottom of the food web (TL = 1, by definition); herbivorous zooplankton feeding on the algae (TL = 2); large zooplankton or small fishes, feeding on the herbivorous zooplankton (TL = 3); large fishes (for example, cod, tuna and groupers) whose food tends to be a mixture of low- and high-TL organisms (TL = 3.5–4.5).

The mean TL of fisheries landings can be used as an index of sustainability in exploited marine ecosystems. Fisheries tend at first to remove large, slower-growing fishes, and thus reduce the mean TL of the fish remaining in an ecosystem. This eventually leads to declining trends of mean TL in the catches extracted from that ecosystem, a process now known as "fishing down marine food webs" (Pauly et al. 1998a) [see Chapter 2].

Declining TL is an effect that occurs within species as well as between species. Most fishes are hatched as tiny larvae that feed on herbivorous zooplankton. At this stage they have a TL of about 3, but this value increases with size, especially in piscivorous species. Because fisheries tend to reduce the size of the fish in an exploited stock, they also reduce their TL.

NATURE

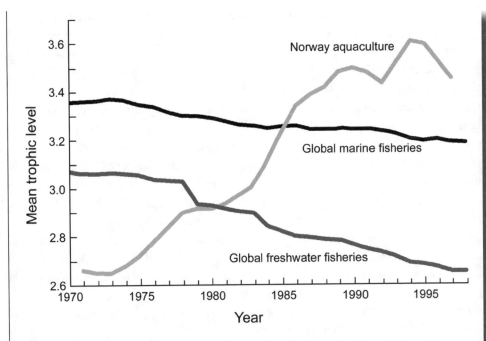

Figure 4.2. Fisheries, both marine and freshwater, are characterized by a decline of the mean trophic level in the landings, implying an increased reliance on organisms low in food webs (data from FishBase, www.fishbase.org, with Peru/Chile excluded owing to the dominance of Peruvian anchoveta; see also Figure 4.1). Freshwater fisheries have lower trophic level values overall, indicating an earlier onset of the "fishing down" phenomenon (Pauly et al. 1998a). The trend is inverted in non-Asian aquaculture, whose production consists increasingly of piscivorous organisms, as illustrated here for Norway, a major producer, yet representative country (Pauly et al. 2001b).

food chains. Rather, predators operate within finely meshed food webs, whose structure (which they help maintain) tends to support the production of their prey. Hence the concept of "beneficial predation," where a predator may have a direct negative impact on its prey, but also an indirect positive effect, by consuming other predators and competitors of the prey (Ulanowicz and Puccia 1990) (and see Box 4.1). Thus, removing predators does not necessarily lead to more of their prey becoming available for humans.[7] Instead, it leads to increases or outbursts of previously suppressed species, often invertebrates (Parsons 1996; Mills 2001; Daskalov 2002), some of which may be exploited (for example, squid or jellyfish, the latter a relatively new resource, exported to East Asia), and some outright noxious (van Dolah et al. 2001).

The principal, direct impact of fishing is that it reduces the abundance of target species. It has often been assumed that this does not impose any direct threat of species extinction as marine fish generally are very fecund and the ocean expanse is wide (Pitcher 1998). But the past few decades have witnessed a growing awareness that fishes can not only be severely depleted, but also be threatened with extinction through overexploitation (Casey and Myers 1998).[8] Among commercially important species, those particularly at risk are species that are highly valued, large and slow to mature, have limited geographical range, and/or have sporadic recruitment (Sadovy 2001). There is actually little support, though, for the general assumption that the most highly fecund marine fish species are less susceptible to overexploitation; rather it seems that this perception is flawed (Hutchings 2000). Fisheries may also change the evolutionary characteristics of populations by selectively removing the larger, fast-growing individuals, and one important research question is whether this induces irreversible changes in the gene pool (Law 2000). Overall, this has implications for research, monitoring and management, and it points to the need for incorporating ecological consideration in fisheries management (Gislason et al. 2000; Hutchings 2000), as exemplified by the development of quantitative guidelines to avoid local extinctions (Punt 2000).

Another worrisome aspect of fishing down marine food webs is that it involves a reduction of the number and length of pathways linking food fishes to the primary producers, and hence a simplification of the food webs. Diversified food webs allow predators to switch between prey as their abundance fluctuates (Stephens and Krebs 1986) and hence to compensate for prey fluctuations induced by environmental fluctuations (Neutel et al. 2002). Fisheries-induced food-web simplification, combined with the drastic fisheries-induced reduction in the number of year classes in predator populations (Longhurst 1998a; Stergiou 2002), makes their reduced biomass strongly dependent of annual recruitment. This leads to increasing variability, and to lack of predictability in population sizes, and hence in predicted catches. The net effect is that it will increasingly look like environmental fluctuations impact strongly on fisheries resources, even where they originally did not. This also will resolve, if in a perverse way, the question of the relative importance of fisheries and environmental variability as the major driver for changes in the abundance of fisheries resources (Cury et al. 2008; Steele 1998; and see Figure 4.3).[9]

It seems unbelievable in retrospect, but there was a time when it was believed that bottom trawling had little detrimental impact, or even a beneficial impact, on the sea bottom that it "ploughed." Recent research shows that the ploughing analogy is inappropriate and that if an analogy is required, it should be that of clear cutting forests in the course of hunting deer. Indeed, the productivity of the benthic organisms at the base of food webs leading to food fishes is seriously impacted by bottom trawling (Hall 1998), as is the survival of their juveniles when deprived of the biogenic bottom structure destroyed by that form of fishing (Turner et al. 1999). Hence, given the extensive coverage of the world's shelf ecosystems by bottom trawling (Watling and Norse 1998), it is not surprising that generally longer-lived, demersal (bottom) fishes have tended to decline faster than shorter-lived, pelagic (open water) fishes, a trend also indicated by changes in the ratio of piscivorous (mainly demersal) to zooplanktivorous (mainly pelagic) fishes (Caddy and Garibaldi 2000).

It is difficult to fully appreciate the extent of the changes to ecosystems that fishing has wrought, given shifting baselines as to what is considered a pristine eco-system (Pauly 1995; Jackson et al. 2001) and continued reliance on single-species models (Box 4.1). These changes, often involving reductions of commercial fish biomasses to a few per cent of their pre-exploitation levels, prevent us taking much guidance from the concept of sustainability, understood as aiming to maintain what we have (Ludwig et al. 1993; Pitcher 2001). Rather, the challenge is rebuilding the stocks in question.

Reducing Fishing Capacity

There is widespread awareness that increases in fishing-fleet capacity represent one of the main threats to the long-term survival of marine capture-fishery resources, and to the fisheries themselves (Weber 1995; Mace 1997). Reasons advanced for the overcapitalization of the world's fisheries include: the open-access nature of many fisheries (Gordon 1954); common-pool fisheries that are managed non-coopera-tively (Munro 1979; Sumaila 1997); sole-ownership fisheries with high discount rates and/or high price-to-cost ratios (Sumaila and Bawumia 2000); the increasing replacement of small-scale fishing vessels with larger ones (Weber 1995); and the payment of subsidies by governments to fishers (Munro and Sumaila 2001), which generate "profits" even when resources are overfished.[10]

This literature shows that fishing overcapacity is likely to build up not only under open access (Clark 1990), but also under all forms of property regimes. Subsidies, which amount to US$2.5 billion for the North Atlantic alone, exacerbate the prob-lems arising from the open access and/or "common pool" aspects of capture fisher-ies, including fisheries with full-fledged property rights (UN 1982; Milazzo 1998).

Even subsidies used for vessel decommissioning schemes can have negative ef-fects. In fact, decommissioning schemes can lead to the intended reduction in fleet size only if vessel owners are consistently caught by surprise by those offering this form of subsidy. As this is an unlikely proposition, decommissioning schemes often end up providing the collaterals that banks require to underwrite fleet moderniza-tions. Additionally, in most cases, it is not the actual vessel that is retired, but its licence. This means that "retired" vessels can still be used to catch species without quota (so-called "under-utilized resources," which are often the prey of species for which there is a quota), or deployed along the coast of some developing country, the access to which may also be subsidized (Kaczynski and Fluharty 2002). Clearly, the decommissioning schemes that will have to be implemented if we are ever to reduce overcapacity will have to address these deficiencies if they are not to end up, as most have so far, in fleet modernization and increased fishing mortality

It is clear that a real, drastic reduction of overcapacity will have to occur if fish-eries are to acquire some semblance of sustainability. The required reductions will have to be strong enough to reduce F by a factor of two or three in some areas, and even more in others. This must involve even greater decreases in f, because catches can be maintained in the face of dwindling biomasses by increasing q (and hence F; see definitions above), even when nominal effort is constant. Indeed, this is the very reason behind the incessant technological innovation in fisheries, which now rely on global positioning systems and detailed maps of the sea bottom to seek out residual fish concentrations previously protected by rough terrain. This technological race, and the resulting increase in q, is also the reason why fishers often remain unaware of their own impacts on the resource they exploit and object so strongly to scientists' claims of reductions in biomass.

If fleet reduction is done properly, it should result in an increase in net benefits ("rent") from the resources, as predicted by the basic theory of bioeconomics (Clark 1990). This can be used, via taxation of the rent gained by the remaining fishers, to ease the transition of those who had to stop fishing. This would contrast with the present situation, where taxes from outside the fisheries sector are used, in form of subsidies, to maintain fishing at levels that are biologically unsustainable, and which ultimately lead to the depletion and collapse of the underlying resources.

Biological Constraints to Fisheries and Aquaculture

Perhaps the strongest factor behind the politicians' use of tax money to subsidize non-sustainable, even destructive fisheries, and its tacit support by the public at large, is the notion that, somehow, the oceans will yield what we need—just because we need it. Indeed, demand projections generated by national and international agencies largely reflect present consumption patterns, which by some means the oceans ought to help us maintain, even if the global human population were to dou-ble again. Although much of the deep ocean is indeed unexplored and "mysterious," we know enough about ocean processes to realize that its productive capacity cannot keep up with an ever-increasing demand for fish.

Just as a tropical scientist might look at the impressive expanse of Canada and assume that this country has boundless potential for agricultural production, unaware that in reality only the thin sliver of land along its southern border (5%) is arable, we terrestrial aliens have assumed that the expanse and depths of the world's oceans will provide for us in the ways that its more familiar coastal fringes have. But this assumption is very wrong. Of the 363 million square kilometres of ocean on this planet, less than 7%—the continental shelves—are shallower than 200 m, and some of this shelf area is covered by ice. Shelves generate the biological production supporting over 90% of global fish catches, the rest consisting of tuna and other oceanic organisms that gather their food from the vast, desert-like expanse of the open oceans.

The overwhelming majority of shelves are now "sheltered" within the exclusive economic zones (EEZ) of maritime countries, which also include all coral reefs and their fisheries (Box 4.3). According to the 1982 United Nations Convention on the Law of the Sea (UN 1982), any country that cannot fully utilize the fisheries resource of its EEZ must make this surplus available to the fleet of other countries. This, along with eagerness for foreign exchange, political pressure (Kaczynski and Fluharty 2002) and illegal fishing (Bray 2000), has led to all of the world's shelves being trawled for bottom fish, purse-seined for pelagic fishes and illuminated to attract and catch squid (to the extent that satellites can map the night time location of fishing fleets as well as that of cities). Overall, about 35% of the primary production on the world's shelves is required to sustain the fisheries (Pauly and Christensen 1995), a figure similar to the human appropriation of terrestrial primary production (Vitousek et al. 1986).

The constraints to fisheries expansion that this implies, combined with the declining catches alluded to above, have led to suggestions that aquaculture should be able to bridge the gap between supply and demand. Indeed, the impressive recent growth of reported aquaculture is often cited as evidence of the potential of that sector to meet the growing demand for fish, or even to "feed the world."

Three lines of argument suggest that this is unlikely. The first is that the rapidly growing global production figures underlying this documented growth are driven to a large extent by the People's Republic of China, which reported 63% of world aquaculture production in 1998. But it is now known that China not only over-reports its marine fisheries catches, but also the production of many other sectors of its economy (Rawski and Xiao 2001). Thus, there is no reason to believe that global aquaculture production in the past decades has risen as much as officially reported.

Second, modern aquaculture practices are largely unsustainable: they consume

Box 4.3

Sustainable Coral Reef Fisheries: An Oxymoron? [*Box originally drafted by D. Zeller*]

Globally, 75% of coral reefs occur in developing countries where human populations are still increasing rapidly. Although coral reefs account for only 0.1% of the world's ocean, their fisheries resources provide tens of millions of people with food and livelihood (Spalding et al. 2001). Yet, their food security, as well as other ecosystem functions they provide, is threatened by various human activities, many of which, including forest and land management, are unrelated to fishing (Roberts et al. 2002). It has often been assumed that the high levels of primary productivity reported for coral reefs imply high fisheries yields (Lewis 1977). However, the long-held notion that coral reef fishes are "fast turnover" species, capable of high productivity, is being increasingly challenged (Choat and Robertson 2002). Yield estimates for coral reefs vary widely, ranging from 0.2 to over 40 tonnes km^{-2} $year^{-1}$ (Russ 1991), depending on what is defined as coral reef area, and as coral reef fishes (Russ 1991; Spalding and Grenfell 1997). Taking yields from the central part of this range (5–15 tonnes km^{-2} $year^{-1}$) and the most comprehensive reef-area estimate available (Spalding et al. 2001), we derive an estimate for total global annual yield of 1.4–4.2 million tonnes. Although these estimates represent only 2–5% of global fisheries catches, they provide an important, almost irreplaceable, source of animal protein to the populations of many developing countries (Russ 1991). Clearly, maintaining the biodiversity that is a characteristic of healthy reefs is the key to maintaining sustainable reef fisheries. Yet coral reefs throughout the world are being degraded rapidly, especially in developing countries (Roberts et al. 2002). Concerns regarding overexploitation of reef fisheries are widespread (Jackson et al. 2001; Roberts et al. 2001; Russ 2002). The entry of new, non-traditional fishers into reef fisheries has led to intense competition and the use of destructive fishing implements, such as explosives and poisons, a process known as "Malthusian overfishing" (Pauly 1994).

Another major problem is the growing international trade for live reef fish (Sadovy and Vincent 2002), often associated with mobile fleets using cyanide fishing, and targeting species that often have limited ranges of movements (Zeller 1997). This leads to serial depletion of large coral reef fishes, notably the humphead wrasse (*Cheilinus undulatus* Labridae), groupers (Serranidae) and snappers (Lutjanidae), and to reefs devastated by the cyanide applications. These fisheries, which destroy the habitat of the species upon which they rely, are inherently unsustainable. It can be expected that they will have to cease operating within a few decades, that is, before warm surface waters and sea-level rise overcome what may be left of the world's coral reefs.[11]

natural resources at a high rate and, because of their intensity, they are extremely vulnerable to the pollution and disease outbreaks they induce. Thus, shrimp aquaculture ventures are in many cases operated as slash-and-burn operations, leaving devastated coastal habitats and human communities in their wake (Pullin et al. 1993; Feigon 2000).

Third, much of what is described as aquaculture, at least in Europe, North America and other parts of the developed world, consists of feedlot operations in which carnivorous fish (mainly salmon, but also various sea bass and other species) are fattened on a diet rich in fish meal and oil. The idea makes commercial sense, as the farmed fish fetch a much higher market price than the fish ground up for fish meal (even though they may consist of species that are consumed by people, such as herring, sardine or mackerels, forming the bulk of the pelagic fishes in Figure 4.1). The point is that operations of this type, which are directed to wealthy consumers, use up much more fish flesh than they produce, and hence cannot replace capture fisheries, especially in developing countries, where very few can afford imported smoked salmon. Indeed, this form of aquaculture represents another source of pressure on wild fish populations (Naylor et al. 2000).

Perspectives

We believe the concept of sustainability upon which most quantitative fisheries management is based (Sainsbury et al. 2000) to be flawed, because there is little point in sustaining stocks whose biomass is but a small fraction of its value at the onset of industrial-scale fishing. Rebuilding of marine systems is needed, and we foresee a practical restoration ecology for the oceans that can take place alongside the extraction of marine resources for human food. Reconciling these apparently dissonant goals provides a major challenge for fisheries ecologists, for the public, for management agencies and for the fishing industry (Cochrane 2000). It is important here to realize that there is no reason to expect marine resources to keep pace with the demand that will result from our growing population, and hopefully, growing incomes in now impoverished parts of the world, although we note that fisheries designed to be sustainable in a world of scarcity may be profitable.

We argued in the beginning of this review that whatever semblance of sustainability fisheries in the past might have had was due to their inability to cover the entire range inhabited by the wildlife species that were exploited, which thus had natural reserves. We further argued that the models used traditionally to assess fisheries, and to set catch limits, tend to require explicit knowledge on stock status and total withdrawal from stocks, that is, knowledge that will inherently remain imprecise and error prone. We also showed that generally overcapitalized fisheries are leading, globally, to the gradual elimination of large, long-lived fishes from marine ecosystems, and their replacement by shorter-lived fishes and invertebrates, operating within food webs that are much simplified and lack their former "buffering" capacity.

If these trends are to be reversed, a huge reduction of fishing effort involving effective decommissioning of a large fraction of the world's fishing fleet will have to be implemented, along with fisheries regulations incorporating a strong form of the precautionary principle. The conceptual elements required for this are in place, for example, in form of the FAO Code of Conduct for Responsible Fisheries (Edeson 1996), but the required political will has been lacking so far, an absence that is becoming more glaring as increasing numbers of fisheries collapse throughout the world, and catches continue to decline.

Given the high level of uncertainty facing the management of fisheries, which induced several collapses, it has been suggested by numerous authors that closing a part of the fishing grounds would prevent overexploitation by setting an upper limit on fishing mortality. Marine protected areas (MPAs), with no-take reserves at their core, combined with a strongly limited effort in the remaining fishable areas, have been shown to have positive effects in helping to rebuild depleted stocks (Mosquera et al. 2000; Murawski et al. 2000; Roberts et al. 2001). In most cases, the successful MPAs were used to protect rather sedentary species, rebuild their biomass, and eventually sustain the fishery outside the reserves by exporting juveniles or adults (Roberts et al. 2001). Although migrating species would not benefit from the local reduction in fishing mortality caused by an MPA (Guénette et al. 2000; Lipcius et al. 2001), the MPA would still help some of these species by rebuilding the complexity of their habitat destroyed by trawling, and thus decrease mortality of their juveniles (Lindholm et al. 1999). Enforcement of the no-take zones within MPAs would benefit from the application of high technology (for example, satellite monitoring of fishing vessels), presently used mainly to increase fishing pressure.

There is still much fear among fisheries scientists, especially in extra-tropical areas, that the export of fish from such reserves would not be sufficient to compensate for the loss of fishing ground (Tupper et al. 2002). Although we agree that marine reserves are no panacea, the present trends in fisheries, combined with the low degree of protection presently afforded (only 0.1% of the world's ocean is effectively protected[12]), virtually guarantee that more fish stocks will collapse, and that these collapses will be attributed to environmental fluctuations or climate change (Figure 4.3). Moreover, many exploited fish populations and eventually fish species will become extinct. MPAs that cover a representative set of marine habitats should help prevent this, just like forest and other natural terrestrial habitats have enabled the survival of wildlife species which agriculture would have otherwise rendered extinct.

NATURE

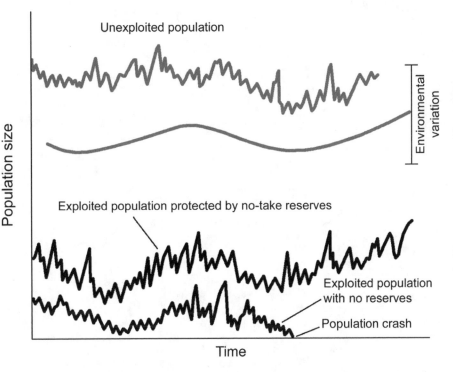

Figure 4.3. Schematic representation of the effects of some environmental variation on an unexploited, exploited but protected, and exploited but unprotected fish population. This illustrates how protection through a marine reserve (and/or stock rebuilding) can mitigate the effects of environmental fluctuations, including "regime shifts" (Steele 1998). Graph from J. Jackson (personal communication).

Focused studies on the appropriate size and location of marine reserves and their combination into networks, given locale-specific oceanographic conditions, should therefore be supported. This will lead to the identification of reserve designs that would optimize export to adjacent fished areas, and which could thus be offered to the affected coastal and fisher communities, whose consent and support will be required to establish marine reserves and restructure the fisheries (Pitcher 2001). The general public could also be involved, through eco-labeling and other market-driven schemes, and through support for conservation-orientated non-government organizations, which can complement the activities of governmental regulatory agencies.

In conclusion, we think that the restoration of marine ecosystems to some state that existed in the past is a logical policy goal (Sumaila et al. 2001). There is still time to achieve this, and for our fisheries to be put on a path towards sustainability.

CHAPTER FIVE

Future of Fisheries

Stepping into the Future

As elaborated below, fisheries scientists do not engage much in crystal ball gazing. Most of their work consists of assembling, every year anew, the evidence needed for estimates of total allowable catches (TAC), which set catch limits for fisheries one year ahead. This is perhaps just as well, as earlier predictions by Thomas Huxley and others about marine fisheries resources' being essentially inexhaustible (Roberts 2007) made it seem prudent to avoid grandiose statements.

There is a way around this, however, and this is by formulating and exploring the consequence of various scenarios. The hosting of a workshop for the Millennium Ecosystem Assessment by the *Sea Around Us* Project in April 2003 was for us a first opportunity to interact with colleagues working on future scenarios regarding human interactions with natural resources (Bennett et al. 2003). Thus in late summer 2003, when the contribution below was solicited by the editor in chief of *Science* for a special series, on the state of the planet, we were well prepared for exploration of societal scenarios, and for embedding fisheries therein.

I drafted the first version of this contribution with Villy Christensen, on his boat, *One Star Shining*, which he uses for both unwinding and exercises such as this. This was then submitted to our coauthors for revision: Elena Bennett emphasized scenario formulation, Jackie Alder matched our scenarios with those of the United Nations Environment Programme (UNEP 2002), Peter Tyedmers concentrated on the energy consumption of fishing fleets, and Reg Watson on the analysis of catch trends, especially those leading to Figure 5.1. Except for some back-and-forth about that figure, the referees were easy to accommodate: there were no requests for change to the text.

SCIENCE

The Future for Fisheries[1]
Pauly et al. | Vol. 302

Formal analyses of long-term global marine fisheries prospects have yet to be performed, because fisheries research focuses on local, species-specific management issues. Extrapolation of present trends implies expansion of bottom fisheries into deeper waters, serious impact on biodiversity, and declining global catches, the last possibly aggravated by fuel cost increases. Examination of four scenarios, covering various societal development choices, suggests that the negative trends now besetting fisheries can be turned around, and their supporting ecosystems rebuilt, at least partly.

Fisheries are commonly perceived as local affairs requiring, in terms of scientific inputs, annual reassessments of species-specific catch quota. Most fisheries scientists are employed by regulatory agencies to generate these quotas, which ideally should make fisheries sustainable and profitable, contributors to employment and, through international trade, to global food security.

This perception of fisheries as local and species-specific, managed to directly benefit the fishers themselves, is conducive neither to global predictions nor the collaborative development of long-term scenarios. Indeed, recent accounts of this type, except those of the United Nations Food and Agriculture Organization (FAO; see note[a]), tend to be self-conscious and layered in irony (Pope 1989; Parrish 1998; Pauly 2000; Cury and Cayré 2001), perhaps an appropriate response to 19th-century notions of inexhaustibility.

The past decade established that fisheries must be viewed as components of a global enterprise, on its way to undermine its supporting ecosystems (Pauly and Christensen 1995; Pauly et al. 1998a; Jackson et al. 2001; Watson and Pauly 2001; Myers and Worm 2003).

These developments occur against a backdrop of fishing industry lobbyists arguing that governments drop troublesome regulations and economists assuming that free markets generate inexhaustibility. The aquaculture sector offers to feed the

[a] The Food and Agriculture Organization of the UN regularly issues demand-driven global projections wherein aquaculture, notably in China, is assumed to compensate for shortfalls, if any, in fisheries landings (see www.fao.org).

[b] Disaggregated global landings assembled by the FAO from 1950–2000 were used to determine when each 30 minute × 30 minute spatial cell was first "fished" (i.e., when landings of fish [other than oceanic tuna and billfishes] from that cell first reached 10% of the maximum landings ever reported from that cell). The percentage of cells fished at each depth was then calculated.

world with farmed fish, while building more coastal feedlots wherein carnivores such as salmon and tuna are fed with other fish (Naylor et al. 2000), the aquatic equivalent of robbing Peter to pay Paul.

The time has come to look at the future of fisheries through (i) identification and extrapolation of fundamental trends and (ii) development and exploration (with or without computer simulation) of possible futures.

The fisheries research community relied, for broad-based analyses, on a data set now shown to be severely biased (Watson and Pauly 2001). First-order correction suggests that rather than increasing, as previously reported, global fisheries landings are declining by about 500,000 metric tons per year from a peak of 80 to 85 million tons in the late 1980s. Because overfishing and habitat degradation are likely to continue, extrapolation may be considered (see below). This correction, however, does not consider discarded "by-catch" (about 30% of global landings), only one component of the illegal, unreported, or unregulated (IUU) catches that recently became part of the international fisheries research agenda (Pauly et al. 2002; Pitcher et al. 2002).

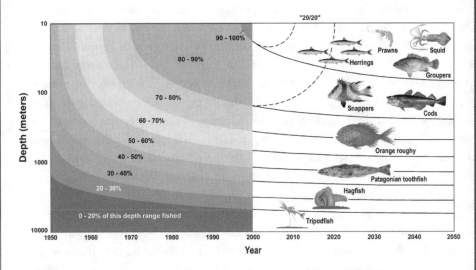

Figure 5.1. Fraction of the sea bottom and adjacent waters contributing to the world's fisheries from 1950 to 2000[b] and projected to 2050 by depth (logarithmic scale). Note the strong reversal of trends required for 20% of the waters down to 100 m depth to be protected from fishing by 2020.

The geographic and depth expansion of fisheries is easier to extrapolate (Figure 5.1). Over the past 50 years, fisheries targeting benthic and bentho-pelagic organisms have covered the shelves surrounding continents and islands down to 200 m, with increasing inroads below 1000 m, whereas fisheries targeting oceanic tuna, billfishes, and their relatives covered the world ocean by the early 1980s (Myers and Worm 2003).

Extrapolating the bottom fisheries trends to 2050 is straightforward (Figure 5.1). With satellite positioning and seafloor-imaging systems, we will deplete deep slopes, canyons, seamounts, and deep-ocean ridges of local accumulations of judiciously renamed bottom fishes, e.g., orange roughy (previously "slimeheads"), Chilean seabass (usually IUU-caught Patagonian toothfish), and hagfish (caught for their "eel-skins," and here predicted to become a delicacy in trendy restaurants, freshly knotted and sautéed in their own slime), the abyssal tripod fishes being the only group that seems safe so far. Figure 5.1 also shows the radical trend change required to turn 20% of the shallowest 100 m of the world ocean into marine reserves by 2020, i.e., returning to the 1970s state.

Traditional explanations of overfishing emphasize the open-access nature of the fisheries "commons." However, overcapitalized fisheries can continue to operate after they have depleted their resource base only through government subsidies (Pauly et al. 2002; Pauly and Maclean 2003). Moreover, industrial fisheries depend upon cheap, seemingly superabundant fossil fuels (Tyedmers 2004), as does agriculture. Thus, we shall here venture a prediction counter to the trends in Figure 5.1, based in part on the global oil production trend in Figure 5.2A: If fuel energy becomes as scarce and expensive in the next decades as suggested by a number of independent geologists (Heinberg 2003), then we should expect the most energy-intensive among industrial fisheries to fold. This would mainly impact deep-sea bottom trawling, which drives the trends in Figure 5.1. One effect may be to increase human consumption of small pelagics (mackerels, herrings, sardines, or anchovies such as the Peruvian anchoveta), now mostly turned into fish meal for agriculture (to grow chickens and pigs, and for use as fertilizer) and aquaculture.[2]

However, predictions are better embedded into scenarios—sets of coherent, plausible stories designed to address complex questions about an uncertain future (Peterson et al. 2003). Scenario analysis is especially important for the fisheries sector, which, although a major provider of food and jobs in many poorer countries, is small relative to the economy of richer countries and is thus "downstream" from most policy decisions.

Pending the detailed analysis of coastal and marine scenarios by the Millennium Ecosystem Assessment (see www.millenniumassessment.org and Bennett et al. 2003),[3] we use the four scenarios developed by the United Nations Environment Programme (UNEP 2002) to investigate the future of marine fisheries. For

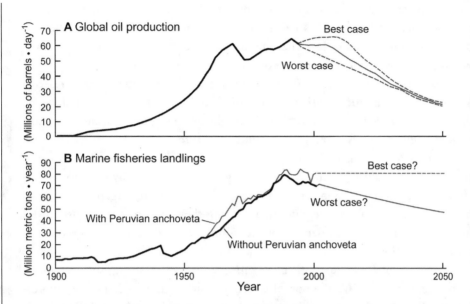

Figure 5.2. Recent historical patterns and near-future predictions of global oil production and fish catches (1900 to 2050). (A) Various authors currently predict global oil production to decline after 2010 (Heinberg 2003), based on M. King Hubbert's model of reservoir depletion, with worst, medium, and best cases based on different assumptions about discoveries of new oil fields. (B) Global marine fisheries landings began to decrease in the late 1980s (Watson and Pauly 2001). The smoothly declining trend extrapolates this to 2050 and also reflects the potential effect of future, exceedingly high fuel prices. The flat line, that is, sustaining present landings, would result from implementing proactive components of the Markets First and Policy First scenarios.

each scenario, we also summarize results of regional simulation models explicitly accounting for interspecies feeding interactions, within a range of ecosystem types and fisheries (see note[c] and Christensen and Walters 2004b).

1) Markets First, where market considerations shape environmental policy. This may imply the gradual elimination of the subsidies fuelling overfishing (Pitcher et al. 2002). Putting markets first may also imply the suppression of IUU fishing (including

---

[c] This work was based on mass-balanced food web models, and their time-dynamic simulation, via coupled differential equations, under the impacts of competing fishing fleets, using the Ecopath with Ecosim software, and models representing the South China Sea (with emphasis on the Gulf of Thailand and Hong Kong waters), the North Atlantic (North Sea, Faeroes), the North Pacific (Prince William Sound, Alaska, and Georgia Strait, BC), and other marine ecosystems documented in www.ecopath.org and www.seaaroundus.org.

flags of convenience), which distorts economic rationality as insider trading or fraudulent accounting does. Markets First, by overcoming subsidies, could also lead to the decommissioning of fuel-guzzling distant-water fleets (especially large trawlers), and perhaps lead to a resurgence of small-scale fleets deploying energy-efficient fixed gears. This scenario allows for spontaneous emergence of quasi-marine reserves (i.e., areas not economically fishable, particularly offshore) and thus may reduce the impact on biodiversity. However, high-priced bluefin tuna, groupers, and other taxa (including invertebrates) would remain under pressure.

When modeled, this scenario corresponds to maximizing long-term fisheries "rent" (ex-vessel values of catch minus fishing costs).[4] This usually leads to combinations of fleets exerting about half the present levels of effort, targeting profitable, mostly small, resilient invertebrates and keeping their predators (large fishes) depressed. Shrimp trawlers presently operate in this way, with tremendous ecological impacts on bottom habitats.

2) Security First, where conflicts and inequality lead to strong socioeconomic boundaries between rich and poor. This scenario, although implying some suppression of IUU fishing, would continue "fishing down marine food webs" (Pauly et al. 1998a), including in the High Arctic, and subsidization of rich countries' fleets to their logical ends, including the collapse of traditional fish stocks. This implies development of alternative fisheries targeting jellyfish and other zooplankton (particularly krill) for direct human consumption and as feed for farmed fish. This scenario, generally accentuating present ("south to north") trading patterns, would largely eliminate fish from the markets of countries still "developing" in 2050.

This scenario would also increase exports of polluting technologies to poorer countries, notably coastal aquaculture and/or fertilization of the open sea. This would have negative impacts on the remaining marine fisheries in the host countries, through harmful algal blooms, diseases, and invasive species.

We simulated this scenario through fleet configurations maximizing long-term gross returns to fisheries (i.e., ex-vessel value of landings plus subsidies, without accounting for fishing costs). The results are increasing fishing effort, stagnating or declining catches, and loss of ecosystem components, i.e., a large impact on biodiversity.

3) Policy First, where a range of actions is undertaken by governments to balance social equity and environmental concerns. This is illustrated by the recent Pew Oceans Commission Report (Pew Oceans Commission 2003), which for the United States, proposes a new Department of the Ocean and regional Ecosystem Councils, and a reform of the Fisheries Management Councils, now run by self-interested parties (Okey 2003).

Similar regulatory reforms, coordinated between countries, combined with marine reserve networks, massive reduction of fishing effort, especially gears that destroy bottom habitat and generate large "bycatch" (Morgan and Chuenpagdee 2003), and abatement of coastal pollution, may bring fisheries back from the brink and reduce the danger of extinction for many species.

This scenario corresponds to simulations where rent is maximized subject to biodiversity constraints. We found no general pattern for the fleet configurations favored under Policy First, because the conceivable policies involve ethical and esthetic values external to the fisheries sector (e.g., shutting down profitable fisheries that kill sea turtles or marine mammals).

4) Sustainability First requires a value system change, favoring environmental sustainability. This scenario, which implies governments' ratification of and adherence to international fisheries management agreements and bottom-up governance of local resources, would involve creating networks of marine reserves and careful monitoring and rebuilding a number of major stocks (Pitcher 2001). This is because high biomasses provide the best safeguard against overestimates of catch quotas and environmental change (Pauly et al. 2002), the latter not covered here but likely to impact future fisheries.

We simulate this scenario by identifying the fishing fleet structure that maximizes the biomass of long-lived organisms in the ecosystem. This requires strong decreases in fishing effort, typically to 20 to 30% of current levels, and a redistribution of remaining effort across trophic levels, from large top predators to small prey species.

These scenarios describe what might happen, not what will come to pass. Still, they can be used to consider what we want for our future. We have noted, however, that many of the fisheries we investigated, e.g., in the North Atlantic (Christensen et al. 2003) or the Gulf of Thailand (Christensen 1998), presently optimize nothing of benefit to society: not rent[4] (taxable through auctions; Macinko and Bromley 2002), and not even gross catches (and hence long-term food and employment security). It is doubtful that they will be around in 2050.

The Editor in Chief's Comments

Donald Kennedy, then editor in chief of *Science*, gathered the papers on the "state of the planet" in a book of the same title (Kennedy et al. 2006). Here is what he wrote about the prospects of fisheries and aquaculture (Pauly et al. 2006) in his introductory chapter (Kennedy 2006):

Marine Resources

Marine resources, especially fisheries, are under at least as much pressure as terrestrial ecosystems. In scoping out the "World Fisheries: The Next 50 Years," Daniel Pauly and a team of colleagues have identified the rates and causes of overharvesting. When open-ocean fisheries were developing in the 19th and early 20th centuries, the dominant view was that these fisheries were essentially inexhaustible. There is something about the opacity of the ocean surface that convinced early fishers and explorers that the inscrutable depths contained untold riches. The rude awakening came after industrialized fishing had been going on for a while following World War II; it was perhaps best embodied in the remark of a fisheries expert that "the modal fate of a North Sea fish is to be eaten by a Scandinavian."

Ocean fisheries constitute a paradigm for the "commons"—a common pool, open-access resource available to a large number of harvesters. The catch of each fisher—call him A—imposes a cost, through depletion of the resource. That cost, however, is shared by all fishers, whereas A enjoys the exclusive benefit of his catch and has an incentive to continue even in the face of declining results. This problem is generalized in the classic "Tragedy of the Commons" [Hardin 1968] and the basis for the phenomena explored in "World Fisheries: The Next 50 Years."

Naturally, the first to experience depletion have been the shallow continental-shelf fisheries that have been historically the most accessible. The groundfish of the New England "banks"—codfish and haddock—were among the first affected, and the plight of those fisheries has presented a chronic political problem for the region. Other in-shore fisheries followed, and the pressure on the industry has gradually moved down in depth and down in the marine food chain, as the illustration in Figure 5.1 demonstrates dramatically. As fishing effort increased globally, total landings went up, but then they flattened out and started to decline in the 1980s.

What goes on as a fishery comes under increased pressure? Declines in the fish population encourage the development of more efficient technology and encourage larger, longer-range, and higher-capacity vessels. That further intensifies the pressure on the resource, so the process has its own positive feedback. Eventually the failure attracts political and economic attention; the first response, unfortunately, is a government subsidy that only provides incentives for the fishers to stay in business.

Better solutions have been offered, and are at work in different settings. Regulations limiting catch, and/or sparing particular size and age classes to preserve reproductive potential can help. Establishing marine reserves has been successful in a number of places. Many advocates for "market-based" solutions favor tradable permits. That system requires solid estimates of the standing population and of catch needed for sustainability; permits are then issued or auctioned to allow that total "take." In such a system efficient fishers catch their take early and buy permits from the less efficient.

Pauly and his colleagues have made an interesting prediction about market forces. The relationship between energy (fuel costs) and the future of fisheries suggests that if costs rose significantly, the most energy-intensive industrial fisheries, particularly the deep sea-bottom trawlers, might fold. The price of crude oil has more than doubled since they wrote, and the prediction will bear watching.

Many of these features are among those employed in the thought-provoking scenarios offered in the chapter by Pauly et al. These provide an excellent example of the complexity of the problem and the diversity of solutions that might be available. It is clear that fisheries differ widely from one another, and that a one-size solution is unlikely to fit all. Since this essay was written, two national commissions—one organized and funded by the Pew [Charitable] Trusts, the other from the U.S. government—have explored solutions for the sustainability of marine resources. The Millennium Ecosystem Assessment also has emerged, with evaluations and recommendations for marine ecosystems.

The special problem of "big fish"— the wide-ranging bluefin tuna, other tuna species, and swordfish—is special in several ways. The incentives for capture are high; in some years an adult bluefin tuna, air-shipped to the Tokyo markets, has brought as much as $30,000. Swordfish—the stars, alas, of *The Perfect Storm*—were for a time a profitable substitute for groundfish for New England fishers. But FDA warnings about mercury as well as warnings about sustainability of the stock have turned off consumers.

In fact the consumer end of the fisheries problem has become a significant project of the environmental movement. Various nongovernmental organizations, university consortia, the Monterey Bay Aquarium, and the David and Lucile Packard Foundation have supported and participated in "Seafood Choice" programs, designed to steer consumers away from fish species with unsustainable populations. Imported swordfish and bluefin tuna have been among the species on the "avoid" list. Public service announcements (for example, "Give Swordfish a Break" and "Pass on Chilean Seabass") have been used, but it will be some time before one can measure their effectiveness.[5]

Aquaculture

On the seafood counter of a local market in Palo Alto, California, there is a striking difference—often as much as a factor of two—between the prices of fish species produced through aquaculture and those caught in the wild by the usual methods. Tilapia, catfish, and rainbow trout are farmed in fresh-water ponds; they are among the least expensive, and are recommended in "seafood choice" programs because they are farmed in environmentally sensitive ways. The case for saltwater aquaculture, especially for salmon and shrimp, is not so good. Salmon are raised in pens that float on the sea surface; escapes are frequent, and the result is often that the gene pool of wild relatives is affected. In order to feed these carnivores, large amounts of fishmeal are required—it takes about 4 pounds of food protein to produce each pound of salmon protein. The growing need for fishmeal is one of the factors encouraging "fishing down the food chain" for less desirable species that can be converted to meal. A study published in 2004 in *Science* demon-strated that some farmed salmon con-tain larger amounts of pesticide residues and other toxicants than wild-caught fish [Hites et al. 2004]. As for shrimp, large aquaculture operations in Asia and elsewhere have done serious dam-age to mangrove shorelines, destroying the nurseries for young wild shrimp and [other] invertebrates. The aquaculture pens have to be changed every few years because the water quality becomes de-graded, and that means that the assault on the mangrove environment is a con-tinuous process. Although it was once thought that aquaculture would pro-vide a rescue from declining traditional fishery yields, it looks as though only the freshwater sector will make a signif-icant contribution to global food needs, though probably not a large one.

The Future Revisited

There is little that needs to be added to this. Clearly, Donald Kennedy "got it," as do most people of goodwill when they are exposed to the facts at hand. We simply made reasonable extrapolations from the best available information. Through the use of di-vergent scenarios, we avoided the error of false precision and put it firmly in the hands of policy makers to choose which ones to make accurate. Thus, there was no need for the fuss that was made by irate colleagues when Worm et al. (2006), extrapolating present trends (i.e., assuming that nothing changes with the way we interact with the ocean[6]), concluded that the overwhelming majority of fisheries will collapse by 2048.[7] We agreed with them when we wrote that "it is doubtful that [these fisheries] will be around in 2050." But it is in our hands to change their course.

Epilogue

Although a first, very incomplete draft of this book was assembled shortly after "The Future for Fisheries" was published, it was only when the final version was completed, in early 2009, that I began to appreciate the profound unity—or consilience (Wilson 1998; Pauly 2002a)—among the five contributions presented therein, despite the wide range of topics covered, and disciplines touched upon. In fact, each contribution calls on the next, and together they form such a clear progression that each step appears straightforward, indeed obvious.

Such obviousness is the topic of Malcolm Gladwell's (2008) neat little book *Outliers*, about persons who have done feats that at first sight appear to be very unlikely, but whose lives offered them many small steps to climb. They climbed them all, finally reaching such heights that they look like outliers.

Outliers, thus, are people lucky enough to have been offered steps to climb (some societies don't offer anything of the sort) and whose main merit is to have taken them, one at a time, thousands of little steps. This is an image I like, although I have climbed far fewer steps than the "outlying" people Gladwell used as his examples.

One can generalize further and note that the little steps Gladwell writes about are similar to those needed to climb "Mount Improbable" (Dawkins 1996), a metaphor for natural selection, discovered by Charles Darwin, one of my heroes (Pauly 2004b), whose 200th birthday we are celebrating as of this writing.

Another characteristic of the five contributions reviewed in this book is that, notwithstanding some critiques, they quickly became widely accepted, as indicated by the number of citations they received.[1] Anecdotal evidence is also available; thus, for example, I have from a reputable source that our contribution of 1995, "Primary Production Required to Sustain Global Fisheries," was among the first decidedly "ecological" papers to penetrate the venerable walls of the Lowestoft fisheries laboratory, and to demonstrate that there was, after all, a relationship between ecology and fisheries. (It was in Lowestoft that much of the mathematical foundation of fisheries science was developed, notably by Beverton and Holt [1957], and it was the leading fisheries institution in the world in the 1960s and 1970s.)

But the best citations, those indicating that your concepts are really getting mainstreamed, are those that you don't get (Garfield 1975), as in "If we continue fishing down the food web it will be more and more damaging for a variety of wildlife species, certainly for the penguins" (Roach 2005). Such non-citation, reflecting what Merton (1968) calls "obliteration by incorporation," means that your work, or a concept that you coined, is seen as so obvious that it can be used without attribution. I see more and more of this, and I appreciate its Zen-like quality, where opposites merge,[2] with the much-cited and the non-cited papers both dissolving in the mighty river of science.

Appendix 1
The Origins of the 100 Million Tonnes Myth

This appendix, adapted from my original response to Baumann's points (Pauly 1996; see Chapter 1), also recaps the old tradition of estimating the potential yields of global fisheries upon which Baumann rested his criticism of our contribution. However, given the contentious nature of the arguments in question, I have split them into numbered steps and documented them by quoting the bits of text that presented the assumptions and numbers crucial to each study (and their sources, if any). No attempt has been made to document all potential yield estimates presented so far (but see Table A1 for a larger sample). Rather, only selected cases are provided, for which, however, all the evidence is included that would be needed to reconstruct the estimate(s) of world ocean potential they contain.

For simplicity's sake, I refer only to marine fisheries; that is, I have omitted fresh-water fisheries, aquaculture (i.e., the farming of fish and other aquatic organisms), and the hunting of marine mammals (whose "potential" is nowadays a political, not a biological, issue). As well, we shall not deal with the details of the earlier work of Moiseev (1969), whose comprehensive study avoided most of the pitfalls identified below but whose English translation of 1971 did not influence the tradition that had then emerged.

The whole story starts at an FAO conference on Fish in Nutrition held in Washington, DC, in 1961, at a time when the world's marine catch was about 35 million tonnes (t) (Figure A1.1). At the time, there was tremendous optimism that the oceans were the source of enough protein to feed the world, and all sorts of schemes were being hatched to develop this apparent potential. At this conference, Graham and Edwards (1962) presented two global estimates of world fisheries potentials (Steps 1–11), using two different methods. In the first, they extrapolated from documented catch levels:

Step 1   "The harvest levels of demersal species, from 7.7 to 12.7 lb. per acre, are surprisingly similar." (This resulted from comparisons among different parts of the North Atlantic and adjacent seas.)

Table A1

Some estimates of the potential fisheries of the oceans

Author(s)	Year	Estimate (t·10⁶ year⁻¹)	Method(s)*	Remarks
Thompson	1951	22	A	—
FAO	1953	55	A, B	—
Finn	1960/1961	50–60	A, B	—
Graham and Edwards	1962	55	B	Bony fishes only, i.e., 2/3 of available nekton
Graham and Edwards	1962	115	C	—
Mesek	1962	55	C	To be reached by 1970
Pike and Spilhaus	1962	180–1400	C	—
Schaefer	1965	200	C	See text
Chapman	1965	2000	C	—
Ricker	1969	150–160	C	Comprehensive review
Ryther	1969	100	C	See text
Gulland	1970	100	A, B	Conventional species only
Gulland	1970	260–350	A, B, C	Including nonconventional species
Idyll	1978	400–700	None	See text
Moiseev	1994	120–150	D	See text

Modified from Schaefer and Alverson (1968) with additions.

*A—extrapolation of catch trends; B—extrapolation from known area to the global ocean; C—extrapolation from primary production and food chain; D—biomass × P/B × factor (as in Moiseev 1994).

Step 2   "The pelagic harvest rates, on the other hand, vary extremely, from 0.2 to 54.2 lb. per acre."

Step 3   "For purposes of computing the possible world harvest from all continental shelves, we can use the value for North Atlantic banks. Let us pick 20 lb. per acre as a conservative average figure" (i.e., 2.2 t km⁻²).

Step 4   "Over the entire globe there are approximately 6×10^9 acres of potentially productive continental shelf" (i.e., 2.4 million km²). Hence, "on this basis, the total continental shelf yield would be 120 billion lb. or 55 million metric tons per year."

Graham and Edwards' first and lower estimate died without progeny; it is their second estimate that became the ancestor of a huge tribe. It was based "on an independent, more theoretical technique" adapted from Kesteven and Holt (1955), as follows:

Step 5   Steeman-Nielsen (1960) estimated the world's primary production as $1.2–1.5 \times 10^{10}$ t·year⁻¹. Assuming a ratio of 37:1 for the carbon to wet weight conversion (Sverdrup et al. 1946) leads to a global annual primary production of about 50×10^{10} t wet weight.

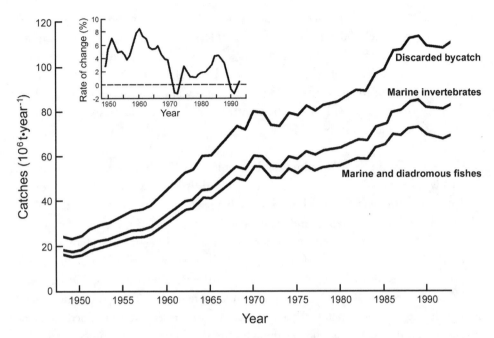

Figure A1.1. Global marine catches, 1948–1993. To account for discarded bycatch, the estimate 27×10^6 t year[-1] of Alverson et al. (1994) was applied to 1992 and prorated to the total catch of all other years. This maintains the structure of the time series, which clearly reflects the collapse of the Peruvian anchoveta in the early 1970s and its increase in the last two to three years, which masks the decline of other major groups. The insert shows the percentage rate of change of the series, smoothed over three years (adapted from FAO Fisheries Statistics yearbooks).

Step 6   Herbivorous zooplankton fully exploit this primary production, and 20% of their food intake is passed on to the next trophic level ("primary carnivores").

Step 7   The primary and secondary carnivores have a transfer efficiency of 10%, leading to an annual global production of 10^{10} t primary and 10^9 t secondary carnivores.

Step 8   About 30% of "theoretical energy transfer" is "sidetracked" (i.e., sedimented and eventually leading to "petroleum deposits"); this may reduce the efficiencies in (6) and (7) to 14%, 7%, and 7%, respectively, and hence "our estimated annual production of secondary carnivores is approximately 343 million metric tons."

Step 9   "Included in this figure, in addition to the marine fish in which we are principally interested are many other marine animals such as squids, whale or sharks which we conservatively estimate make up one-third of the consuming biomass at this and higher consumer levels. All things considered, then perhaps 230 million metric tons of marine bony fishes are produced on an annual basis."

Step 10   "Properly harvested, it is reasonable to suggest that resources of this nature may yield 50 per cent by weight, at least, of the net annual production."

Step 11   "We estimate finally, therefore, that perhaps as much as 115 million metric tons of marine fishes may be available for harvest each year. A significant part of this resource is thinly scattered [... and] our fishery technology will have to be greatly improved before a harvest of this magnitude is possible or feasible."

Of Graham and Edwards' estimates, the first is based on data (Steps 1–4), although heavily weighted by the assumptions that the North Atlantic was then fully exploited, and globally representative. The line of reasoning leading to the second estimate, while also starting with data (Steps 5 and 6), depends entirely on the guesses in Steps 6–10, of which 10 is the most interesting: we shall encounter it again, if under a different guise. It is the first mention of such a 50-50 split between harvest and predation in the literature, but, as we shall see, it was to live on, although it was nothing but a guess.

In science, experimental findings are considered reliable if, using the same general approach but other, independently derived data, other researchers obtained the same results (i.e., if they could "replicate" these findings). However, these other researchers can also obtain the same findings if they make the same, unexamined assumptions, leading to the same errors. We will see that this was the case when multiple authors took up the challenge implicit in Graham and Edwards' findings.

Schaefer (1965) identified three approaches for estimating potential yields: (a) "by extrapolation of recent trends (a dangerous business for looking more than a few years into the future)," (b) "by considering our knowledge of unused harvestable resources," and (c) "by calculations based on food chain dynamics." He then presented a fine review of earlier attempts to estimate marine potential yields and concluded by presenting his estimate, which relied entirely on (c). His line of reasoning was as follows:

Step 12   Slobodkin (1961, p. 138), based on laboratory experiments on water fleas (*Daphnia* spp.) in the lab, estimated growth efficiency at "around 8.5 to 12–13 percent."

Schaefer took this as equivalent to a transfer efficiency of about 10% between trophic levels; thus, he continued:

Step 13   "Effective ecological efficiency, due to [...] recycling, may be higher than the 10% estimate; 15% would not seem an unreasonable guess and 20% should be possible."

Step 14   "Currently, 37% of the marine fishing harvest consists of [fish that] feed on a mixture of phytoplankton and zooplankton. So, perhaps this harvest corresponds to about 1 1/2 steps above the phyto[plankton]. The remainder of the harvest is from levels a step or two higher. We are, I think, very conservative if we assume the harvest is all taken at [trophic level] 3."

Step 15   Given a primary production of 19×10^9 t carbon (from Pike and Spilhaus 1962, and "based on very inadequate data"), a 10% efficiency between trophic levels leads to a potential of 190×10^6 t; with 15%, this leads to 640×10^6 t per year.

Step 16   Moreover, "if we assume that half of the potential might be taken at [trophic level 3] and half of [trophic level 4], which is more nearly realistic," the available potentials become 1080 and 2420 million t per year, respectively.

Step 17   "Only a part of this can be realized, because of economic inability to harvest some of the components that are diffusely distributed, and because other predators than man take a share of the potential harvest."

Step 18   "A minimum estimate of 200 10^6 tons would appear to me reasonable and probably conservative."

Step 19   "Calculations of this sort by Graham and Edwards provide somewhat similar results [...]. If one includes the other organisms, discarded in Graham and Edwards' calculation, the corresponding estimate of available harvest is 171×10^6 tons."

Hence, Schaefer's line of reasoning starts with data (Step 12); all subsequent steps are based on guesses; the last step then shows that the result is similar to that obtained in an earlier study, itself based on a series of guesses.

Ryther (1969) appears to have been among the first to use geographic strata or "provinces" for estimating world fisheries potential: (a) the open seas (326 million km²) with a primary production of 50g C m⁻² year⁻¹ and a food chain length of 5 steps (PP; herbivores; first-, second-, and third-stage carnivores, e.g., tuna); (b) coastal waters (i.e., areas down to 100 fathoms, or 180 m, slightly less than the 200 m limit commonly used to define continental shelves) and "offshore regions of comparably high productivity," with a mean primary production of 100 C m⁻² year⁻¹, a surface area of 36 million km², and a food chain length of 3 steps; (c) upwelling systems, with an approximate area of 0.36 million km², a mean primary production of 300 g C m⁻² year⁻¹, and a food chain length of 2 1/2 steps (primary producers, herbivorous zooplankton, and fish feeding on both phyto- and zooplankton). The transfer efficiencies required for each stratum were then guessed (or more precisely, were guessed based on Slobodkin's data on growth and feeding experiments with water fleas, just as Schaefer's were), and the conclusions drawn:

Step 20   "Slobodkin (1961) concludes that an ecological efficiency of about 10% is possible and Schaeffer [sic] feels that the figure may be as high as 20%.

Here therefore, I assign efficiencies of 10, 15 and 20 percent, respectively, to the oceanic, the coastal and the upwelling provinces, though it is quite possible that the actual values are considerably lower." [And indeed they are, particularly for upwellings where the transfer efficiency is 4–8% (Figure 1.5), i.e., 2.5 to 5 times the value guessed by Ryther.]

Step 21   "In all, I estimate that some 240 million tons (fresh weight) of fish are produced annually in the sea. As this figure is rough and subject to numerous sources of errors, it should not be considered significantly different from Schaeffer's figure of 200 million tons."

Step 22   "Production however is not equivalent to potential harvest. In the first place, man must share the production with other top level carnivores [...]. In addition, man must take care to leave a large enough fraction of the annual production to permit utilization of the resource at something close to its maximum sustainable yield [...]. When these various factors are taken into account, it seems unlikely that the potential sustained yield of fish to man is appreciably greater than 100 million tons."

Once again, Graham and Edwards' work is "validated." Particularly fascinating is the fact that Ryther's conclusion, like Graham and Edwards' and Schaefer's, relies on a 50-50 split between fisheries and predators, which no longer merely defines the amount "available" but now estimates the likely "potential sustained yield" or even the fabled "maximum sustainable yield."

Alverson et al. (1970), who suggested that Ryther may be "right, but for the wrong reasons," took issue with almost all of Ryther's numbers, assumptions, and illustrative examples (not documented here). Notably, they suggested that "his selection, for his calculations, of relatively high trophic levels from the coastal and oceanic provinces is questionable." As Box 1.2 illustrates, this point is crucial. However, Alverson et al. (1970) did not comment on the high transfer efficiencies Ryther used, for which he presented even less justification than for the trophic levels (though a subsequent study shall "confirm" them; see below).

Alverson et al. (1970) also mention a then ongoing, major FAO study (which led to the report edited by Gulland in 1970) that, they implied, would lead to more reliable estimates than Ryther's approach. As we shall see, that study did lead to a potential yield (of conventional species) close to Ryther's estimate and thus it was "right" (but again, for the wrong reason).

The lucid review of Gulland (1970), building on the then established tradition, proposed three approaches for global potential estimates:

- extrapolation of catch trends;
- extrapolation from known areas to the global ocean; and
- extrapolations from primary production and transfer efficiencies.

Gulland then went on to discuss the pitfalls of these methods, and likely errors associated with each. His subsequent estimate of global potential yield differs from those listed above in that the logic of its derivation cannot be summarized by a short sequence of pithy statements. Rather, his overall estimate, that is, the global potential for "conventional" species, of 100 million t year^{-1}, results from the addition of a large number of largely independent, area-specific estimates obtained by various authors commissioned by FAO through various combinations of methods (a), (b), and (c). This approach, similar to that of Moiseev (1969), leads to estimates that are far more robust than the estimates obtained by multiplication of a few guessed numbers.

To account for "unconventional" species, Gulland (1970) estimated annual catch potentials of 10–100·million t for cephalopods, "100+" [million t] for "lanternfish, etc.," and "50+" [million t] "for euphausiids in Antarctica."

One important feature of the FAO review edited by John Gulland, and widely circulated in book form (Gulland 1971), is that he presented in its preface the derivation of the (in)famous "Gulland equation"—whose true originators appear to be Alverson and Pereyra (1969). This equation relies, for estimation of potential yield, on a logic similar to that first presented by Graham and Edwards (1962) and adopted by subsequent authors (see above), but not cited in Gulland's preface. It has the form "Potential yield = $0.5 \cdot M \cdot B_0$" where M is the natural mortality of the stock in question, and B_0 its unexploited biomass. This was based on two sets of arguments:

Step 23 The surplus production model of Schaefer (1954) implies that unexploited biomass is halved when maximum sustainable yield (MSY) is achieved. Thus, if the fishing mortality generating "MSY" (F_{MSY}) is about equal to natural mortality (M), then the logic in Gulland's equation applies.

Step 24 Yield-per-recruit (Beverton and Holt 1964) is optimized for values of mean lengths at first capture ranging from 40 to 70% of asymptotic length, when fishing mortality is about equal to natural mortality, or $F_{opt} \approx M$.

Given that production/biomass ratio (P/B) is equivalent to total mortality ($Z = F + M$) for standard representations of growth and mortality (Allen 1971), Gulland's equation implies that MSY represents 50% of biological production—precisely the assumption of Graham and Edwards (1962; see above).

Many subsequent authors built on this assumption (see, e.g., Dickie 1972; Parsons and Chen 1994), but dedicated studies (Francis 1974; Beddington and Cooke 1983; Kirkwood et al. 1994; Christensen 1996) show it to be untenable: the fishing mortality that maximizes sustainable yield is for most single- or multispecies fish stocks much smaller than M, or $F_{MSY} \approx 0.2$ to $0.5 \cdot M$. The implications for potential yield estimates that assume $F_{MSY} \approx M$ are obvious (you will overestimate potential catch!). Also, as noted by L. Alverson (personal communication), it is not obvious that M at B_0 is equal to M at $B_0/2$.

Nonetheless, based on his review of many of the papers cited above, Idyll (1978) concluded:

Step 25   "[T]he biological evidence is strong that the potential for familiar kind of seafood is 100 to 120 millions tons."

Step 26   However, impressed by Gulland's estimates for unconventional species, he concluded that "it does not seem unreasonable to suppose that a total of all species of 400 millions tons could be caught, and it might be as large as 700 million tons."

Iverson (1990) did not set out to estimate the global potential of the ocean but, perhaps more ambitiously, to identify the factors that "control [...] marine fish production." His methodology and data sets are rather opaque and need not concern us here except insofar as he deals with trophic levels (TL) and the transfer efficiencies between these, which he relates through the following steps:

Step 27   The first was that "Fish production = Primary production·(transfer efficiency)^{TL}" where the trophic level is "set equal to a non-integer value to represent fish production as the average of production on several trophic levels (Ryther 1969)."

Step 28   "The results of an analysis of N stable isotope data (Fry 1988) suggest that the average trophic status of the species providing most of Georges Bank fish production can be characterized by [a mean trophic level] = 2.5. A similar value was assumed for food chains on northeast North American coastal environments [...]; catches of the Baltic Sea environments include herring, sprat and cod, which are assumed to be characterized by an average value [of trophic level] similar to the Georges Bank value."

These guesses, plus a few more for mesopelagic fishes, and a bit of fiddling led to the curve in Figure A1.2, which Baumann (1995) cites, as it "confirmed" the guessed transfer efficiencies assumed by Ryther (1969) and, before him, by Schaefer (1965). However, because the guesses were wildly off the mark, I am confident that these "results" were obtained by backward calibration.

Steps 1–28 illustrate how a scientific community can become fixated upon assumption and procedures that stop being questioned, and which tend to be "confirmed," even if the data have to be bent in the process.

This becomes obvious when contrasting independent analyses are available. Thus, in a brief communication presented in 1992 at the first World Fisheries Congress, Moiseev (1994) included a table whose ten elements add up to a global fish biomass of 6–7 billion t, and an annual production of 4–5 billion t, much higher than the above estimates. Further, he suggested:

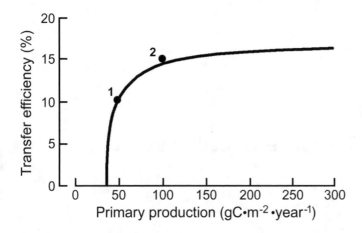

Figure A1.2. Reproduction of Figure 7 in Iverson (1990), showing putative "transfer efficiency as a function of total phytoplankton production. [...] Estimate of transfer efficiency for (1) oceanic and (2) coastal non-upwelling environments were assumed by Ryther, 1969." Note that this neat "confirmation" fails, however, to accommodate Ryther's 20% guess for upwelling systems (see text and Figure 1.5).

Step 29   "The biomass of large fish and other organisms which form the basis of conventional present day fisheries is estimated at about 1.5 billions tonnes." Using a P/B ratio of 0.4 year^{-1}, he estimates for these a production of 600×10^6 t year^{-1}.

Step 30   Moiseev then concludes that "the possible yield of conventional fishing may reach 120–150 million tonnes (20–25%) of the production; that of organisms at lower trophic levels may be many times more."

Thus, Moiseev, working within an intellectual context independent from that then prevailing in the West, assumed that only "20–25%" of fish production (P = biomass × P/B) could be sustainably exploited, rather than 50%. He started, however, with a bigger biomass and hence arrived, overall, at a figure resembling recent catches (Pauly et al. 2002). We tend to take more than 20–25% of the annual production of "conventional fish." Hence, if we trust Moiseev's figure, we are catching too much for our fisheries to be sustainable.

Appendix 2
Rejoinder: Response to Caddy et al.[a]

D. Pauley, R. Froese, and V. Christensen | *Science* 282 (5393)

Our report (Pauly et al. 1998a) received a lot of media attention, some of it over-enthusiastic. Thus, we are pleased that Caddy and his colleagues at FAO have provided, through their detailed comment, an opportunity to elaborate on the process of fishing down marine food webs, wherein fishing fleets actively and increasingly target species low in the food web. Caddy et al. seem to agree with us when they state that "a general decline in mean trophic level of marine landings is likely to have occurred in many regions." They question, however, whether FAO landing data can be used to demonstrate the existence of this trend.

We would like to clear up two possible misunderstandings before we turn to the four objections made by Caddy et al. First, we did not state, or imply, that low trophic level species increased their contribution to global catches because of a "depletion of their predators." Rather, we suggested that continuation of the trend to fish down marine food webs must eventually lead to declines of overall catches (both predator and prey species), resulting in "backward-bending" curves of trophic level against these catches [see Figure 2.6]. We identified several mechanisms that could generate such curves, including one in which the removal of top predators reduces the production of their prey. This mechanism is different from the "predator depletion" model.

Second, values produced by the mass balance ("Ecopath") ecosystem models, from which we extracted the more than 200 estimates of trophic level used to compute mean trophic levels of FAO landings, were not values that were "assigned," that is, input into Ecopath, but values that were estimated by Ecopath, on the basis of observed diet compositions. We now turn to the four considerations raised by Caddy et al.

[a] Contribution originally authored by D. Pauly, R. Froese and V. Christensen; submitted August 19, 1998; accepted September 15, 1998, and published online November 20, 1998. Reprinted with permission.

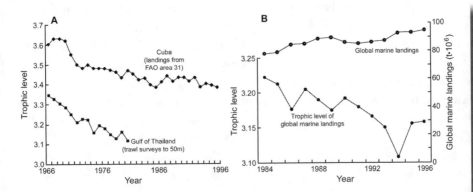

Figure A2.1. Trophic level trends (A) in Cuban landings from the western central Atlantic (FAO Area 31, 1966–1996) and in Gulf of Thailand trawl survey data, 1966–1982 (Christensen 1998); (B) in global marine fisheries, 1984–1996 (from FAO landings, also shown), after removal of mariculture production data. (The catch trends in B do not correct for overreporting by China; see Chapter 3.)

(i) The lack of "taxonomic resolution" is indeed a problem in the FAO landing data set. However, we demonstrated global and broad regional trends toward lower trophic level in spite of about half of the world's landings being assigned to excessively broad categories, such as "mixed fishes." This is especially true in tropical developing countries, and as fishing down marine food webs also occurs in these countries [Figure A2.1, panel A], the overall effect is actually much stronger than we were originally able to show. FishBase 98 (Froese and Pauly 1998) may be used to generate graphs similar to that for Cuba in [Figure A2.1, panel A] for a vast array of countries, all with similar trends, even where over aggregated regional data do not exhibit fishing down marine food webs. As a rule, we find that the better the taxonomic resolution, the stronger the effect of fishing down marine food webs appears.

(ii) Using "landing data as ecosystem indicators" is not really a problem: landings of major resource species should generally reflect the relative magnitudes of their biomasses in the ecosystems from which the landings are extracted. Thus, Peruvian landings consist mainly of anchoveta because these are abundant in the Peruvian upwelling ecosystem, and Indonesian coastal fishers land ponyfishes because these are abundant on the Sunda Shelf. Off Newfoundland, Canada, where cod was targeted until it recently collapsed, a fishery for invertebrates has recently developed. It can be safely expected that Newfoundland's future landing statistics will reflect the species shift that occurred in the ecosystem around that island.

Such correspondence between relative abundance in the landing and in the eco-systems was not the rule before fisheries became globalized, and only selected species were exploited by nearshore gear. Now, with inshore, offshore- and distant-water fleets competing to supply increasingly integrated global markets, abundant spe-cies are exploited wherever they occur (Grainger and Garcia 1996), and landings will tend to reflect their relative abundance. Moreover, there is evidence for fishing down marine food webs in fisheries, independent of FAO data (Christensen 1998). One example is the nearly two decades of well-documented surveys in the Gulf of Thailand (Pauly 1988) [Figure 2.8A]. There, fishing down marine food webs cannot be shown when using highly aggregated, regional FAO data (Pauly et al. 1998a). Also, given the strong positive relationship, in aquatic ecosystems, between trophic level and size (1), both within and between species [Figure A2.1], the occurrence of fishing down marine food webs implies a reduction of mean size for the exploited components of aquatic ecosystems. Reduction of mean sizes in multispecies fisheries catches (commercial and surveys) are themselves very well documented in the litera-ture (Rice and Gislason 1996). Caddy et al. state that we did not consider within-species (that is, ontogenic) changes of trophic level. (We admit having planned to leave this for another paper [i.e., Pauly et al. 2001a]). In fish, trophic level does not simply "change" during the transition from larvae to adults: it increases [Fig-ure A2.2, panel B]. Because most species of fish, globally, "have seen a significant increase in their exploitation resulting from the spread of new technologies," their mean size, and thus their mean trophic level, cannot but have declined in recent years. Our not considering, in the report, within-species changes of trophic level masked the full extent of fishing down marine food webs, instead of artificially cre-ating it, as implied by Caddy et al.

(iii) Aquaculture development is another issue we had reserved for a later contribution. The trend of trophic level in global aquaculture is the oppo-site of that in fishing down marine food webs: it is increasingly carnivorous, high-trophic-level species (salmon, groupers) that are cultivated, while the low trophic-level (herbivorous and detritivorous) species popular in developing countries (tilapia, carp) are either phased out or grown for sale to upscale mar-kets, using fish meal or other high-protein diets. Similarly, the transition from wild-caught (largely detritivorous) penaeid shrimps to shrimp culture, relying on high-protein pelleted feeds, also implies a trophic level increase.

Caddy et al. attempt to demonstrate [Figure 2.7] that the inclusion of aqua-culture production in FAO landing data may have produced, rather than masked, the decline of trophic level we reported. We re-analyzed the FAO database, after

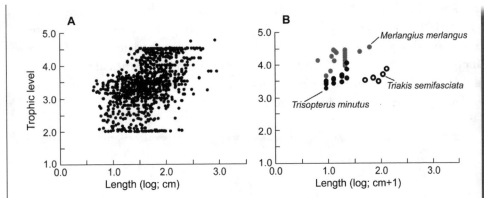

Figure A2.2. Relationships between trophic level and body length in fish. (A) Trophic level from diet compositions versus maximum length in 1143 species. (B) Trophic level from diet compositions versus mean predator length in three representative species. All data are from FishBase 98 (Froese and Pauly 1998).

excluding freshwater fishes, non-fish vertebrates such as whales, and algae and other plants as we had done before—and also excluding aquaculture production from 1984 on (to reflect when the FAO Aquaculture Production database started). The result shows the same decline as reported earlier, about 0.1 trophic level unit per decade (Pauly et al. 1998a). Indeed, a decline of trophic level also occurs in [Figure 2.7] of the comment by Caddy et al., even for their 1984–1996 series, presenting landings minus aquaculture production, although this trend is barely visible, due to the inappropriate scale of [Figure 2.7].

(iv) Eutrophication of coastal areas may have caused a "bottom up" effect in increasing abundance of planktivores, thus lowering mean trophic levels. We agree that this effect may be one of the causes for some of the observed declines in the trophic level of fisheries landings—if we assume that, indeed, changes in abundance in the ecosystem tend to be reflected in the landings. This, however, is a point Caddy et al. did not grant us in other parts of their comment.

Currently, eutrophication is limited to coastal areas, including parts of the Mediterranean. In the Black Sea and along the Louisiana Coast (Gulf of Mexico), the existing fisheries have become shadows of their former selves because of overfishing, extreme eutrophication, and other anthropogenic disturbances. This is likely also to have happened in other areas similarly affected, but on the global level, catches from such areas are not significant, and thus will have only minor impact on trends of trophic level. Caddy and his colleagues have documented, and tried to halt, the

SCIENCE

excessive global fishing capacity that has depleted major fisheries (Grainger and Garcia 1996). We, and they, have relied on the vast effort that went into generating and maintaining the global FAO database of landings and related data. We have supported this effort and shown in our report how combining the contents of this database with the knowledge derived from ecosystem models can provide further insight into what is happening globally.

Appendix 3
Post-1998 Studies of "Fishing Down"

The following text, mainly based on material originally assembled by Villy Christensen, lists and briefly comments on some contributions demonstrating the occurrence of "fishing down" using local, taxonomically disaggregated data sets, following the original presentation of this phenomenon for broad regions, and globally, by Pauly et al. (1998a). The studies are presented north to south by continent, then for the world ocean.

Africa

Senegal and Guinea, northwest Africa (Laurans et al. 2004): The study, covering the years 1970 to 2000, was based on the analysis of catch and trawl survey data, the latter providing an estimate of ecosystem biomass; landings showed a stronger trophic level (TL) decline.

Ebrié Lagoon, Côte d'Ivoire (Albaret and Laë 2003): "[T]he response of the Ebrié lagoon fish community to high fishing pressure is to evolve toward a smaller number of species, especially planktophag[ous] or detritivor[ous] ones."

Namibia, northern Benguela system (Willemse and Pauly 2004a, 2004b): This is one of the world's ecosystems most affected by overfishing, and in which most indicators have been declining since the 1970s (total catch; the FiB index, and an index of piscivory, i.e., the faction of piscivorous fishes in the catch over the total catch; see Caddy and Garibaldi 2000), but in which the trend in mean trophic index appears to be overly sensitive to the choice of TL for a few key species. However, the main reason why the mean trophic level (of the catch) fails to decline is that the groups that are now overwhelmingly abundant in that ecosystem—pelagic goby and jellyfish (Lyman et al. 2006)—are not presently targeted by any fishery.

America (North)

Western Canada (Pauly et al. 2001a), based on a comprehensive data set assembled by Wallace (1999), and covering the years 1873–1996: There has been overall decline of mean TL since 1900, with some fluctuations when new fisheries were initiated.

Eastern Canada (Pauly et al. 2001a), based on data submitted to FAO by Canada's Department of Fisheries and Oceans, and covering the years 1950–1997: There has been a clear decline (by 1 TL from 1955 to 1995), which becomes stronger when accounting for the within-species size (and hence trophic-level) reduction that accompanies fishing down.

Quoddy Region, Bay of Fundy, eastern Canada (Lotze and Milewski 2004): Fishing down has been shown to be part of multiple damages, accelerating in recent decades.

Gulf of Maine (Steneck et al. 2004): Three phases are identified for the trophic structure of the Gulf of Maine: "Phase 1 is characterized by vertebrate apex predators such as Atlantic cod, haddock, and wolffish and persisted for more than 4,000 years. Phase 2 is characterized by herbivorous sea urchins and lasted from the 1970s to the 1990s. Phase 3 is dominated by invertebrate predators such as large crabs and has developed since 1995. Each phase change resulted directly or indirectly from fisheries-induced 'trophic-level dysfunction' in which populations of functionally important species at higher trophic levels fell below the densities necessary to limit prey populations at lower trophic levels."

Chesapeake Bay and US Mid-Atlantic (Chuenpagdee et al. 2006): The catches from Chesapeake Bay, the US Mid-Atlantic coast, and FAO Area 21 (northeastern Atlantic) were shown to exhibit declining and converging mean TLs; Chesapeake Bay and the US Mid-Atlantic coast started in 1950, at much lower TL values (2.6–2.8) than the more encompassing FAO Area 21, which started at a mean TL of 3.4.

St. Augustine, Florida (Reitz 2004): "Comparing zooarcheological data for Native American, Spanish, and British occupations with modern fisheries data from St. Johns County, Florida (USA) shows difference in the use of marine resources from 1450 B.C. through A.D. 2000. Changes in biomass contribution, diversity, types of fishes used, and trophic levels of sharks, rays, and bony fishes suggest that the pattern described as 'fishing down marine food webs' (Pauly et al. 1998a) may have been present in the St. Johns County area as early as the eighteenth century."

Gulf of California (Sala et al. 2004): "Coastal food webs in the Gulf of California have been 'fished down' during the last 30 years—fisheries shifted from large, long-lived species belonging to high trophic levels to small short-lived species from lower trophic levels. In addition, the maximum individual length of the landings has decreased about 45 cm in only 20 years."

Cuban EEZ (Baisre 2000), covering the years 1960–1995: Clear downward trends of mean trophic level and mean maximum size were obtained, which "indicate a gradual transition from landings of large piscivorous fish to small fishes and invertebrates feeding on smaller organisms. These results are in line with those obtained by Pauly et al. (1998a) on a global scale."

Western central Atlantic (Pauly and Palomares 2005), based on data for FAO Area 41, covering the years 1950–2000, disaggregated into USA (north) and other countries (south): "Fishing down" became visible after the masking effects of spatial overaggregation were corrected for (see Figure 2.9).

Caribbean (Wing and Wing 2001): These authors studied archaeological sites "on five Caribbean islands, each with an early (1,850–1,280 years B.P.) and late (1,415–560 years B.P.) occupation. On each of these islands (Puerto Rico, St. Thomas, St. Martin, Saba, and Nevis), the mean size of reef fishes in the faunal remains declined from the early to the late occupation. [...] The mean trophic level of reef fishes declined from the early to the late occupations on each island. Together these patterns suggest that populations of reef fishes adjacent to occupation sites on these islands were heavily exploited in prehistoric times."

America (South)

Peru/Chile (Pauly and Palomares 2005): There is a clear pattern of declining mean TL (1950 to late 1990s) once the immense catches of Peruvian anchoveta (*Engraulis ringens*) are removed from the catches from FAO Area 87 (southeastern Pacific). The remaining catch consists essentially of coastal demersal species, and hence the diagnosis of "fishing down" applies to the coastal/benthic ecosystem.

Chile, central (Arancibia and Neira 2005): "Because total landings off Central Chile have been strongly influenced by landings of horse mackerel (*Trachurus symmetricus*), which is a trans-zonal fishery resource (*sensu* FAO), and the periodic occurrence of El Niño Southern Oscillation (ENSO) events in the study area, we explore changes in [mean trophic level] TL_m both excluding landings of horse mackerel and excluding landings in ENSO years. When landings of horse mackerel are excluded, a significant decline in TL_m is found, at a rate of 0.175 per decade, which is higher than the global rate of 0.10 estimated by Pauly et al. (1998a). Fisheries in Central Chile seem to have been fishing down the food web as the result of fishery-induced changes, since ENSO events do not seem to have induced a significant effect in this trend. Therefore, we suggest that landings of horse mackerel have masked the fishing down the food web process in local fisheries."

Uruguay (Milessi et al. 2005): "A decline in total landings (Y) is observed, which is explained by the lower fishing yield in major fishery resources (especially demersal). Moreover, a marked decreasing trend in TLM at a rate of approximately 0.28 trophic levels per decade, and a decreasing trend in FiB-index since 1997 were observed. The present situation of fishery resources in Uruguay (fully exploited or overexploited) and the drop in Y, FiB and TL_m can be considered as indirect indicators of the fishing impacts on the trophic structure of Uruguayan marine communities."

Asia

India (east and west coasts), and Indian States (IS) and Union Territories (UT) (Vivekanandan et al. 2005; Bhathal and Pauly 2008): The former study showed declining TLs only for the Indian west coast, while the latter, working with spatially dissagregated data, showed such decline to occur in each single IS and UT (see Figure 2.10).

Gulf of Thailand (Christensen 1998; Pauly and Chuenpagdee 2003): The former study demonstrated a clear TL decline of the trawlable biomass from 1963 to 1982, while the latter showed decline of mean TL of catch data covering 1977 to 1997.

China, coastal waters (Pang and Pauly 2001): Decline of the mean TL was observable, despite the low resolution and overall reliability of catch data, which covered the years 1950 to 1998.

Europe

Iceland (Valtysson and Pauly 2003), based on data set in Valtysson (2001), and covering the years 1900–1999: There was a decline from a plateau of mean TL = 3.8–3.9 in the period from 1915–1935 to around 3.4 in the 1990s; this decline increases when one considers within-species changes in size and hence trophic level.

North Sea (Heath 2005), based on data for 1973 to 2000: "The secondary production demand per unit fish production synthesizes trophic guilds and species composition changes that have occurred at least since the 1970s, and hence may be a useful index of foodweb fluxes in the North Sea. Declining values of this index have reflected the loss of, in particular, demersal piscivores, and indicate a case of 'fishing down the food web', in the terminology of Pauly et al. (1998a)."

Celtic Sea (Pinnegar et al. 2002), based on trophic levels estimated from stable nitrogen isotopes and catch and survey data covering the years 1945–1998: "Our analysis showed that there has been a significant decline in the mean trophic level of survey catches from 1982 to 2000 and a decline in the trophic level of landings from 1946 to 1998. The decline in mean trophic level through time resulted from a reduction in the abundance of large piscivorous fishes and increase in smaller pelagic species which feed at a lower trophic level."

Cantabrian Shelf, Spain (Sanchez and Olaso 2004): "The mean trophic level of Cantabrian Sea fisheries declined from 1983 to 1993 but has remained steady since then."

Portugal, 1927–1999 (Coelho 2000): Mean trophic levels increase from 3.0–3.1 in the late 1920s to 3.4 in the late 1960s, then decline to 3.15–3.20 in the late 1990s. These changes, however, appear mainly to reflect the relative abundance of sardines.

Venice Lagoon, 1945–2001 (Libralato et al. 2004): "The time series of landings in the Venice Lagoon from 1945 to 2001 were analysed with the aim of explaining the ecosystem changes [that] occurred. The comparative analysis of the total landings and mean trophic level (mTL) time series allowed to identify four different stages in the lagoon ecosystem. The first period, from 1945 to 1973, was characterised by increasing trends in the landings and their mTL. The second one, from 1974 to 1989, showed a decrease in the landings but still an increase in the mTL. The third period, from 1990 to 1998, had again a positive trend in the landings, but the mTL showed a sharp decline. After 1998, a slight decreasing trend in both mTL and landings was observed."

Greek waters (Stergiou 2005): Fishing down was stronger for higher "TL-slices" (see the section FAO's Comments and a Rejoinder, in Chapter 2); see also Stergiou and Konlouris (2000).

World

Sibert et al. (2006), based on tuna data covering the years 1950 to 2000, suggested, "The trophic level of the catch dropped from 4.1 to 4.0 over the past 50 years because of the increased catch of smaller fish." On the other hand, the small but steady decline of mean trophic level in Figure 2 (top) of Pauly and Palomares (2005), suggesting "fishing down" within the global community of large pelagic fishes (ISSCAAP Group 36), was an artifact due to the unrealistically low trophic level (3.7) erroneously assigned to yellowfin tuna (*Thunnus albacares*; see Table 1 in Pauly and Palomares 2005), whose catches have increased in recent decades, both absolutely and in relation to those of other large pelagics. However, the increase of the FiB index in Figure 2 (bottom) of these same authors, suggesting a massive spatial expansion of the fisheries for large pelagic fishes, holds. Moreover, this expansion should not be viewed as having occurred only over a surface area (of ocean), but as also involving successively deeper water layers. It should be straightforward to redefine the FiB index such that it allows for a volume-based interpretation (see Box 2.2), but developing a case study may be arduous.

Pauly and Watson (2005) presented two global views of fishing down, one as a bivariate graph (mean TL vs. time), and the other as a map of TL changes between 1950 and the present. Jointly, they indicate that fishing down is a strong and ubiquitous effect.

Endnotes

Preface

1. Most of the articles in question can be downloaded from the Web site of the *Sea Around Us* Project (www.seaaroundus.org) and/or from my personal Web site (www.fisheries.ubc.ca/members/dpauly/).

2. My experience with this process is not limited to the ideas presented in the five contributions featured in this book. My work in the tropics, mainly structured around the analysis of length frequency data using the newly developed ELEFAN methods, also experienced (in high-latitude countries) a vehement initial resistance, as did the further development and dissemination of the Ecopath approach and software (see note 23 in Chapter 1) and the development of FishBase (see Box 2.1). The furious debates and fiery publications that resulted are now a matter of the past (and thus need not be documented here, which is a pity, as some of this stuff was quite funny), and ELEFAN, Ecopath, and FishBase are now widely accepted. However, these earlier episodes undoubtedly influenced, for better or for worse, my responses to the reception of the five contributions covered here, and hence the shape of this book.

Chapter 1

1. The start of my work at the University of British Columbia (Fall 1994) coincided with a decision to become more directly involved in conservation issues and with environmental nongovernment organizations (NGOs). My previous work was focused on developing techniques for fish stock assessments tailored to the specificities of the tropics (review in Pauly 1998) and disseminating them through courses on all continents (notably courses I gave on behalf of FAO; Venema et al. 1988). The resulting exposure to similar problems in many different countries gradually convinced me that the results of fisheries scientists and the management advice based thereon, in the tropics and elsewhere, are generally ignored when fleet owners and/or governments make decisions about investment in fisheries (including port development) or when fleets are deployed. Thus, to protect fisheries from their own suicidal tendencies (and to protect the biodiversity and ecosystems that they depend on), outside pressure must be applied. Such pressure is best generated by civil society, i.e., by conservation-oriented or, generally, by environmental NGOs.

2. This is now the WorldFish Center, based in Penang, Malaysia (see www.worldfishcenter.org).

3. I was born in Paris, France, but grew up in the French-speaking part of Switzerland. At 17, I left that country for Germany, where I completed high school through evening courses, and then went to study at the University of Kiel, on the shore of the Baltic Sea. I had intended to eventually work in Africa (Malakoff 2002) and did the fieldwork for my master's thesis in Ghana (see Pauly 1975). However, I ended up for almost two years in Indonesia, and when I completed my doctoral studies—again at Kiel University—the dice had been cast for a return to Southeast Asia, this time in the Philippines.

4. A good example is provided by length-based population growth and stock assessment models, which are age-structured in temperate regions (because the age of commercial fishes can be readily estimated), while they are length-based in tropical regions (where the ages of fish can be estimated only with considerable difficulties, if at all) (Pauly 1998).

5. See Ward (2004) for an account of the philosophy behind the massive data collection and dissemination by the United Nations and its specialized agencies, notably the FAO. The latter organization began issuing its indispensable *Yearbooks of Fishery Statistics—Catches and Landings* in 1950, which is the reason why most time series in this book can start as early as they do. I salute the visionaries who set up this global system right after the Second World War.

6. The Soviet Union was then the "East," as China had not burst onto the world scene.

7. Actually, it turned out that they couldn't (Walters and Maguire 1996), but that is another story.

8. This is what the central limit theorem is all about; see, e.g., Sokal and Rohlf (1995).

9. Transfer efficiency (also known as trophic efficiency or ecological efficiency) is estimated by dividing the flux of biomass (or energy) at trophic level $n + 1$ by that at trophic level n. Ideally, one's estimate of transfer efficiency should be a mean of many single estimates. Transfer efficiency within ecosystems is sometimes estimated from the growth efficiency of single organisms, i.e., by dividing their growth (or better, their "production"; Winberg 1971; Pauly 1986a) in a given time interval by their food consumption during that same interval, but this leads to biased estimates of transfer efficiencies. Finally, the term *conversion efficiency* is used occasionally but should not be, because it does make a distinction between organisms and ecosystems.

10. Here, and throughout this book, I replaced the system of numbered endnotes that *Nature*, *Science*, and some other journals use by the more common "Harvard system," where references are indicated by author and date, thus enabling readers to identify cited papers in their proper context. The endnotes, instead, are used for comments and additional information.

11. This statement actually anticipates ideas presented in Chapter 2.

12. Also, at the time the MSY concept emerged, it provided a rationale for limiting fishing effort, and for delayed gratification ("catch less now so that you will catch more later"). Moreover, its numerical value could be calculated (for single species), albeit by using increasingly complex methods. This is why the concept still survives, despite its "epitaph" having been written (Larkin 1977). Right now, a catchy concept that would capture something like

"maximum (?) sustainable ecosystem yield" does not exist. In fact, estimates of MSY (or one of its variants) that explicitly account for predation by other animals in the system are still extremely rare.

13. Contribution originally authored by D. Pauly and V. Christensen; manuscript received August 24, 1994; accepted January 26, and published March 16, 1995. Reprinted with permission.

14. The word *tonnes* (abbreviated "t") is used throughout this book to represent 1000 kg, or one "metric ton."

15. Note that ellipses in square brackets identify, here and elsewhere in this book, deletions inconsequential to the case being made.

16. No, we did not.

17. Actually, I should have written "asymptotic length," a technical term referring to the mean length the fish of a given population would reach if they were to grow forever. This, and the yield per recruit concept, are detailed in Pauly (1998) and in the classics of fish population dynamics, notably Beverton and Holt (1957).

18. The price that we pay for overfishing has been estimated at about half of the ex-vessel value of the fisheries of the world, i.e., about US$50 billion per year (World Bank 2009), far in excess of the uneducated guess by Lomborg (2001; see following note).

19. Many of those who reviewed Lomborg (2001) noted that it covers too many environmental disciplines for a single person to provide a competent review of the whole oeuvre. Therefore, they concentrated on their areas of expertise (global warming, species extinction, water supply, etc.) and found that, at least for those areas, Lomborg's Panglossian zeal had driven him to quote selectively from the literature and/or to misrepresent it (see www .lomborg-errors.dk/). This is also true for my review (Pauly 2002b), from which the rest of this note is adapted.

The *Skeptical Environmentalist* devotes only two pages to fisheries and aquaculture (p. 106–108, plus a few scattered mentions on other pages), in line with its author's belief, based on his use of misleading global averages, that "fish constitute a vanishingly small part of our total calorie consumption—less than 1 percent—and only 6 percent of our protein intake stems from fish."

Lomborg admits that fisheries have an overfishing problem, however, "[t]he oceans could produce about 100 million tons of fish a year, which we can harvest 'for free' (in the sense that we do not have to feed them). Right now, we only catch about 90 million tons, the missing 10 million tons being the price we pay for over-fishing the sea [FAO 1997a]. Not catching the extra 10 million tons is inefficient, but in effect equivalent to just putting the world food development back a bit less than three weeks."

Note the disconcertingly cavalier attitude about what clearly is a bigger issue than setting back "world food development"—whatever that is—by three weeks. There are many countries in the world whose inhabitants rely to a large extent on fish for their food and/or income. Whole-world consumption figures are entirely inadequate for dealing with issues of food security.

But then, Lomborg feels he does not really need to care, since "total fish production has increased so much that the fish per capita in the late 1990s once again exceeded all previous years," citing FAO's *State of the World Fisheries and Aquaculture 2000* ("SOFIA"; FAO 2001a) as his source. The irony is that this was the last issue of SOFIA to claim that world fisheries catch increased through the 1990s. The subsequent SOFIA, published in 2002, took account of the fact that China had massively overreported fisheries catches in the 1990s, generating in the process an increase later shown to be spurious (see Chapter 3).

We cannot blame Lomborg for having taken these misleading statistics at face value; he is not the only one who did. But we should fault him for posing as an expert, then telling us that stagnating or declining trends of fish landing per capita do not matter anyway, because we can always farm the fish we need, i.e., "it appears of minor importance whether the consumer's salmon stems from the Atlantic ocean or a fish farm." You can feed flippancy to your readers, but you must feed fish to salmon, in the form of fish meal and oils (farmed salmon otherwise get sick and die, or survive and taste like tofu), and this fish has to come from some fishery (Naylor et al. 2000; Alder et al. 2008). Yet Lomborg had conceded that catches cannot be increased above 100 million tons (see above). We will leave it at this.

20. An overview of tuna catch distributions may be found, e.g., in Fonteneau (1997).

21. See also note 8.

22. Another example of this is the "Drake equation" used to estimate the number of extraterrestrial civilizations (say in our galaxy) based on the number of stars, the number of potentially life-bearing planets per star, the fraction of such planets with some forms of life, etc. Some versions of the Drake equation lead to high estimates of the number of extraterrestrial civilizations, which give lots of weight to Fermi's question when confronted with these results: "Where is everybody?"

23. The Ecopath software, originally developed in Hawaii by Jeff Polovina and colleagues (Polovina and Ow 1983, 1985; Atkinson and Grigg 1984; Polovina 1984a, 1984b) for routine construction and validation of food web models of ecosystems, was further developed at ICLARM, first under my leadership (see Pauly et al. 1993), then under that of Villy Christensen, and the resulting product has been well documented (Christensen and Pauly 1992a, 1992b) and widely applied (see, e.g., the models in Christensen and Pauly 1993a, Pauly and Christensen 1996, and Christensen and Maclean 2004). In the 1990s, I and later Villy Christensen moved on to the Fisheries Centre of the University of British Columbia and immediately began collaborating with Carl Walters, who applied his formidable mathematical and programming skills to the extension of Ecopath into a time-dynamic model (Ecosim; Walters et al. 1997) and a spatially explicit model (Ecospace; Walters et al. 1999). These three modules now complement each other and can be used for cross-validation (Pauly et al. 2000).

24. In 2008, the US National Oceanic and Atmospheric Administration listed on its Web site (www.noaa.gov) Ecopath among its 10 greatest accomplishments ever (in over 100 years, if you include the times when it was not called NOAA), crediting J.J. Polovina for its creation, and Villy Christensen, Carl Walters, and me for its further development and dissemination. Irwin et al. (2003) shows how far it got.

25. Two research developments contribute, nowadays, to the widening of the applications of the concept of primary production required (PPR) by fisheries. One of these is a nifty indicator of ecosystem "health," which builds on earlier work by Tudela et al. (2005) and

Libralato et al. (2008) and which accounts for both PPR and the catch (of potential predators) foregone by catching prey fish (Coll et al. 2008).

The other new development is the extension of the "footprint" concept to the ocean realm. Footprint analysis, invented by Rees and Wackernagel (1994) as a conceptual tool enabling comparisons of the impact of various human activities on the Earth's ecosystems, essentially consists of expressing these activities in terms of the surface area required for generating products or for absorbing the waste generated in the course of supplying these products. Various conversion tables exist for such standardized analyses, e.g., for crop production, or the absorption of carbon emissions and other waste products (Wackernagel and Rees 1996). Also important to footprints is that, generally, they are expressed in relative terms, e.g., in terms of the surface area of a country. Thus, a country that has a footprint exceeding its surface area relies on resources from other countries. The footprint concept and the conversion tables that are used to implement it are, however, tied to land areas.

One way to extend the footprint concept to the marine realm is to relate the PPR by the fisheries of a maritime country to the primary production in its exclusive economic zone (EEZ), i.e., the area of the ocean to which it has exclusive use within the constraints of the international Convention on the Law of the Sea (UNCLOS 1994). This is implemented, for each maritime country, by the *Sea Around Us* Project (www.seaaroundus.org), enabling comparisons of the relative ecosystem impacts of fisheries on the EEZs of different countries.

However, this does not consider landlocked countries, which also have, through their import and consumption of seafood, an ecological impact on the oceans. This last point—and more—can be accommodated straightforwardly by adding the primary production required locally by all seafood consumption (fisheries catches + mariculture products) to that required for all seafood imports. This is the basis of the "seafoodprint" concept, which is at the heart of the collaborative project initiated in late 2008 between the *Sea Around Us* Project and *National Geographic*. Clearly, primary production required has come a long way.

Chapter 2

1. See note 2, Chapter 1.

2. Rigler (1975) argued that trophic levels were an empty, nonquantitative and, hence, nonscientific concept and thus expressed a line of arguments that began with ecosystem research (Golley 1993), or at least since trophic levels were proposed as a device to sort animals into (Lindeman 1942). These arguments stemmed from the difficulty some ecologists had in reconciling this concept, articulated in the form of integers (primary producers = 1; first-order consumer = 2, etc.), with the (correct) observation that many consumers are omnivores, i.e., they feed on more than one trophic level.

However, this problem was overcome by the introduction, through Odum and Heald (1975), of fractional trophic levels, computed as weighted means from diet composition data. Also, the fact that such trophic levels (and their variance) can be computed and compared with estimates derived from stable isotope ratios (Kline and Pauly 1998) indicates that, far from being only an unquantifiable "concept," trophic levels are properties of animal populations and ecosystems, just as, e.g., temperature is.

3. Essentially, this routine assigns fractions (p_i) of the observed stomach contents of predators (e.g., p = 0.60 anchovies, 0.20 swimming crabs, 0.15 shrimps, and 0.05 unidentified matter) to preset categories (i), each of which is assumed to have a certain trophic level (TL_i), estimated from previous studies. The trophic level of the predator (j) is then computed from $TL_j = 1 + \sum p_i \cdot TL_i$. It should be mentioned that trophic levels can also be estimated from the ratio of ^{14}N to ^{15}N isotopes (Minagawa and Wada 1984; Pinnegar et al. 2002). Comparison of estimates shows that the two methods give similar results (Kline and Pauly 1998).

4. This International Conference on the Sustainable Contribution of Fisheries to Food Security, held in Kyoto in December 1995, was one of many conferences sponsored by Japan to promote the study of predation in marine ecosystems for the purpose of creating a positive environment for its whaling activities. Villy Christensen's paper at that conference demonstrated, however, that the overwhelming bulk of predation in marine ecosystems consists of fish preying on other fish, and that hence removing marine mammals generally will not increase fisheries yield (Christensen 1996).

5. In fact, this 10% efficiency value pops up everywhere one looks, including in our food (see Morowitz 1992).

6. Contribution originally authored by D. Pauly, V. Christensen, J. Dalsgaard, R. Froese, and F. Torres Jr.; manuscript received August 22, accepted December 10, 1997, and published February 6, 1998. Reprinted with permission.

7. Mass-balance models (e.g., Ecopath models, see note 23 from Chapter 1) are so named because the sum of the energy or biomass flows entering the model's compartments (i.e., the "functional groups," or "boxes," or "state variables") must by definition be equal to the sum of flows leaving those same compartments—at least after a set period (conventionally a year). This requirement is usually met by natural ecosystems, which tend to remain self-similar over time.

8. Contribution authored by J.F. Caddy, J. Csirke, S.M. Garcia, and R.J.R. Grainger; submitted July 24, 1998; accepted September 15, 1998, and published online November 20, 1998. Reprinted with permission.

9. Figure 2.7 in the comment by Caddy et al. (1998b) has several problems. It is based on 39 highly aggregated trophic level estimates from an earlier contribution (Pauly and Christensen 1995), rather than on the more than 200 estimates of trophic level used in Pauly et al. (1998a), and made available through www.fishbase.org. Further, it suggests trophic level values for aquaculture as a whole to be increasing from 1.55 in 1984 to about 1.7 to 1.8 in the 1990s. Such low trophic levels require that mariculture production of plants (trophic level = 1) be high relative to that of herbivores/detritivores (trophic level = 2), and carnivores such as salmon (trophic level >> 2, because their diet includes animal products, such as fish meal). Last, we do not see what algae and seagrasses, explicitly excluded from the computations in Pauly et al. (1998a), can contribute to a debate about fishing down marine food webs, except to generate an inappropriate ordinate scale for graphs of trophic level over time (see Tufte 1983) [This note was originally part of our rejoinder to Caddy et al., now Appendix 2.]

10. As we shall see in Chapter 3, throughout the 1990s, global marine fisheries catches were actually decreasing.

11. The first thing that we had to set straight in our first response (Appendix 2) was the contention of Caddy et al. that we stated or implied that low-trophic-level species increased their contribution to global catches because of a "depletion of their predators." Rather, we suggested that continuation of the trend to fish down marine food webs must eventually lead to declines of *overall* catches (both predator and prey species). They were also inaccurate in stating that we drew "global conclusions about the effects of fishing on world fish stocks with the use of research data fitted to Ecopath models at different sites through the world's oceans, integrated with data on global fishery landings collected by the [FAO]." In reality, the only role that Ecopath models had in this research was to provide some of the 200+ estimates of trophic levels we used.

12. Within-species changes of (size-specific) trophic levels were also considered in Valtysson and Pauly (2003), in their analysis of fishing down in Icelandic waters, with similar results.

13. I now believe that identifying the appropriate level of spatial aggregation is crucial to detecting occurrences of fishing down. For example, tuna and other large pelagics with an overwhelmingly oceanic distribution must be omitted when studying fishing down in neritic, or continental shelf, ecosystems (Pauly and Palomares 2005; Bhathal and Pauly 2008). Indeed, on similar grounds, we should have shown as Figure 2.2A the global trend of mean trophic levels with and without Peruvian anchoveta (as we later did for landings in Figure 3.3). This would have generated a clearer signal (Figure 4.2) and perhaps led to an earlier acceptance of our estimate of a global trophic level decline of 0.05 to 0.10 per decade.

Appendix 3 includes most of the studies that, at this time of writing, have been published and have explicitly addressed "fishing down" as defined in Pauly et al. (1998a). Thus, they can be seen as tests of the hypothesis that fishing down is widespread. Declining mean trophic levels can be seen as a local indicator of sustainability, if only for the simple reason that any downward or upward trend cannot go on for a long time.

On the other hand, the absence of a downward trend of mean trophic level does not demonstrate that the fisheries in question are sustainable, because the fishing-down effect can be masked by various biasing effects (see the "judo arguments" in the main text). Rather, sustainability can be asserted only after examination of a number of indicators, not only the mean trophic level of the catch.

14. "Farming up" the food web is the process wherein the farming of low-trophic-level organisms (e.g., oysters, mussels) is replaced by the farming of carnivorous fish (e.g., salmon, sea bass, tuna) fed fish meal or small fishes (Pauly et al. 2001b; Stergiou et al. 2009).

15. Fishing "through" vs. "down" the food web is probably a distinction without much of a difference, given that, other things being equal, larger, longer-lived fish with higher trophic levels will generally be more strongly affected by a mixed fishery (e.g., trawling) than small, shorter-lived fish with lower trophic levels (Cheung et al. 2005). Thus fishing "through" the food web is at best a transient phenomenon, a stage of the more general fishing down. Indeed, we wrote in the first paragraph of the original contribution that "[f]ishing down leads at first to increasing catch, then to a phase transition associated with declining catches" (Pauly et al. 1998a, and see above). The very fact that every year FAO adds catches from dozens of new species in its landings database while landings *decrease* (FAO 2009) is

sufficient evidence that fishing "through" lacks generality (besides being directly refuted in specific cases, as shown further down in the main text, and in Appendix 3).

Lack of generality is also an issue with Litzow and Urban (2009) who, after they demonstrated that trophic levels in Alaska have not declined in the last five decades (they did decline from earlier levels when salmon made the bulk of the catch, but this is beside the point), concluded that, hence, fishing down is not as widespread globally as asserted in many papers that I was associated with (see also Appendix 3). However, they should have emphasized, instead, that fishing down, although it is widespread, does not occur in Alaska, because the exploitation rate on most stocks therein is low. Or put differently: either fishing down is not widespread, and its absence from Alaska does not indicate anything, or fishing down is widespread, and its absence from Alaska shows how well Alaskan fisheries are managed, in contrast to almost everywhere else. But they can't have it both ways.

16. To assist in this multinational research project, the Web site of the *Sea Around Us* Project (www.seaaroundus.org) now includes a page for each country (even for the USA, although it's not a member of the CBD), with the MTI (and the FiB index, see above) for the fisheries occurring in its EEZ, with a facility to select the years and species included in the analysis (Pauly 2005). This facility is also available for large marine ecosystems (LMEs) and was used extensively to help establish the status of the 66 LMEs recognized as of this writing (Sherman and Hempel 2008; Pauly et al. 2008).

17. This Science section has now become a Science and Health section, reflecting a trend affecting many other newspapers. This is sad, if you think about it.

18. Except that I was renamed "David" Pauly.

19. Also rewarding was that my daughter, then at the International School Manila, was assigned this *New York Times* article as the topic for a term paper by a geography teacher who did not know why she beamed upon receiving the assignment.

20. The invitation was issued by William "Monty" Graham, whom I met later, with his graduate students, at the Dauphin Island Marine Laboratory, on the Alabama coast of the northern Gulf of Mexico. This meeting, my first with jellyfish people, led to our collaborating on a paper (Pauly et al. 2008), which described how "fishing down marine food webs" tends to create conditions favorable to jellyfish outbursts.

21. The "Myxocene" (from the Greek μγχο—mucus, slime) is here proposed as an alternative to the self-congratulatory "Anthropocene."

Chapter 3

1. Richard Grainger's presence at this event illustrates, incidentally, that scientific arguments—such as that between myself and Caddy et al. (Richard was one of the *alia*)—usually do not spill over into other areas, and hence continued collaboration is possible. This is not the case when *ad hominem* arguments are used, which did not occur here.

2. The *Sea Around Us* Project was designed to document in the scientific literature and disseminate to a wide audience information about global fisheries impacts on marine ecosystems and policies that would mitigate such impacts (Pauly 2007). Maps, because of their potentially huge information contents and their ability to explain complex phenomena (think of weather maps, which are extremely information-rich but which everybody understands),

are very important to our goal (see also Pauly and Pitcher 2000), and we make hundreds of them available on our Web site (www.seaaroundus.org).

Data-rich *global* maps are very important in this context, because they can reveal patterns that local maps may not resolve. This is similar to our emphasis on long time series, typically back to 1950, which can reveal patterns (i.e., trends) that shorter time series cannot see. Thus, the *Sea Around Us* Project can be perceived as analogous to the bigger machines (particle accelerators, telescopes, etc.) that physicists build when they cannot resolve a phenomenon (or an issue) with the machines they have.

The only problem with our global maps and long time series is that we always get in trouble with colleagues who work with 10 years worth of local data. This is similar to the situation prevailing with *global* warming, which the ignorant think is invalidated by the cold wind we had last week in Sète.

3. This is restated in the long response to our contribution further in the main text—but you have to read it carefully.

4. The "zero growth policy" was 100% successful: after 1998, catch (actually, *catch reports*) remained at exactly the 1998 level (see Figure 3.1B). In fact, FAO (2009) still reports from China "a very stable capture production." The folks at *Saturday Night Live* could not have made this up.

5. This offers great opportunities to bring Sherlock Holmes into the story, whom I already tried to recruit for the *Sea Around Us* Project, for the very purpose of clearing up misleading catch statistics (Pauly 2003). We already have the Dr. Watson that is required for exclaiming "elementary, my dear Watson!"

6. Thurow's argument is two-pronged: (a) China's electricity usage is reported to have grown at about 5% per year, which is half the reported increase of the overall economy. Given that electricity is used in all sectors of the economy, it is doubtful that the overall economy grew 10% per year over decades. (b) Given that Chinese economic growth appears to be in urban areas, with about 30% of the country's population, it would have to grow at a rate of 33% in the coastal areas to result in a 10% national average growth—which is hardly conceivable.

7. The propensity to inflate output figures can mask and amplify the effects of famines, as happened in China from 1958 to 1961. Chang (1991, p. 224–225) gives an account of such inflation: "In many places, people who refused to boast of massive increases in output were beaten up until they gave in. In Yibin, some leaders of production units were trussed up with their arms behind their backs in the village square while questions were hurled at them:

'How much wheat can you produce per *mu*?'

'Four hundred *jin*' (about 450 pounds—a realistic amount).

Then beating him:

'How much wheat can you produce per *mu*?'

'Eight hundred *jin*.'

"Even this impossible figure was not enough. The unfortunate man would be beaten, or simply left hanging, until he finally said: 'Ten thousand *jin*.' Sometimes, the man died hanging there because he refused to increase the figure, or simply before he could raise the figure high enough."

8. These catch maps, which have now become a major product of the *Sea Around Us* Project (Watson et al. 2001b, 2004), enable us, for example, to illustrate trends at the level of countries' exclusive economic zones (EEZs) and large marine ecosystems (LMEs), both of which allow inferences on the impact of fisheries on ecosystems (see www.seaaroundus.org or Sherman and Hempel 2008).

9. About 50% of all submissions to *Nature* are turned down without peer review. Of those sent to reviewers, about 20% obtain the unanimously positive reviews required for a submission to be accepted, yielding an overall rejection rate of 90%. These numbers are similar for *Science*.

10. He told me later. It is not what you thought, whatever it was.

11. As of this writing, Jane Lubchenco has been confirmed as Undersecretary of Commerce in the Obama administration, in charge of the National Oceanic and Atmospheric Administration and hence of it subagency, NOAA Fisheries, earlier known as National Marine Fisheries Service (NMFS). Congratulations, Jane!

12. Contribution originally authored by R. Watson and D. Pauly; manuscript received July 30, accepted September 28, and published November 29, 2001. Reprinted with permission.

13. They should have a rough time these days, when most people realize that these very tenets caused the financial and moral crisis we are in.

14. This is actually "Organization." Can the Brits change names just like that?

15. Here, a few references would have added much to the argument. . . .

16. This may sound strange, but I have met many colleagues in developing countries who did not know more about their countries' fisheries than what is found on FAO's Web site— one more reason for these data to be as reliable as possible.

17. As we shall see, this is also what our contribution achieved.

18. This very long response by FAO could have been reduced to this short sentence. In fact, this would have been the best response to the media onslaught: "China is working with FAO to try to address these issues."

19. Note, however, that the report by Alverson et al. (1994), which documented immense amounts of discarded bycatch, was not integrated in the FAO catch statistics, even though it was published by FAO.

20. This "slow process" could be accelerated, however, by FAO's opening itself to academia and civil society (i.e., environmental NGOs), both of which could contribute to a more open discussion on the creation, content, and validity of these data sets, which, after all, are a collective property of humankind, rather than the effective property of FAO staff (as one could assume in view of their defensiveness when independent analyses of these data are performed). Indeed, the independent sources of funding that these new actors could tap may help mitigate the fact that "financial support for the development and maintenance of national fishery statistical systems has declined sharply in real terms."

Presently, however, FAO is still doing what it perceives as damage control, e.g., when they suggest, when discussing the state of world fisheries, that "[t]hese management failures have given rise to widespread concerns, often accompanied by high profile media reports, about the negative impacts of fisheries on marine ecosystems. In the eyes of many environmentalists and of public opinion in general, the overfishing of stocks, habitat modification resulting from destructive fishing practices, the incidental capture of endangered species and other impacts have made fisheries a primary culprit in an ecological crisis of global dimensions. While some of the claims have been exaggerated and some are misleading, the underlying crisis is real and an urgent response is required at global level. However, in responding, there is a danger that the pendulum will swing too far in the other direction and, from an over-emphasis on short-term social and economic goals, the long-term goal of conservation will become the only driving force in the management of human impacts on aquatic ecosystems" (FAO 2009, p. 36).

Actually, if any clear message is to be extracted from this prose, it is that FAO staff are still happy, in 2009, to contrast "short-term social and economic goals" (i.e., the ability of the owners of fishing fleets to make a quick buck) and the maintenance of the ecosystems from which humans extract their sustenance—as if this mattered only to "environmentalists."

21. The Code of Conduct for Responsible Fisheries (FAO 1995) is a voluntary scheme to which countries and private actors may or may not comply (Edeson 1996). T.J. Pitcher, at the UBC Fisheries Centre, and his collaborators have been evaluating the performance of over 50 countries, who jointly contribute about 95% of the world catch, on implementing this code (Pitcher et al. 2006, 2009a, 2009b). Their major finding was that "overall, compliance with the Code was dismal" (T.J. Pitcher, personal communication, February 3, 2009).

22. The terms *ecosystem-based management* and the related *ecosystem-based fisheries management* warrant some comments: In the 1970s, when I was a student of fisheries biology in Germany, we were taught from textbooks, by R. Beverton and S. Holt, J. Gulland, W. Ricker, and others, that did not mention more than casually the ecosystems within which the exploited species in question are embedded. In fact, these species were presented as having mainly internal dynamics, upon which the "outside," i.e., the ecosystem, could impact only via natural mortality, which was represented by a constant ("M"). It was the same in other countries, and there were perhaps good reasons for such reductionism (see Pauly 1990). However, two related developments led to these simplifications' gradually becoming unacceptable, however useful they had been at first. One was the inordinate number of exploited stocks that crashed in the 1980s and 1990s, the most famous of these being that of northern cod off eastern Canada, which was supposedly well managed and whose collapse hugely impacted on the credibility of fisheries science (see also note 29 below). The other development was the collateral damage of fisheries, in the forms of drowned marine mammals, seabirds, and sea turtles and habitat destruction, both of which emerged as big public issues in that period.

I do not know who first coined the term *ecosystem-based fisheries management* (EBFM), which was proposed as the solution to these ills. But the term was extremely successful, as attested by its acceptance by the NGO community and even the public, often before governmental and intergovernmental fisheries management agencies could adapt. In fact, the quick acceptance of the term preceded serious scientific discussion of its implications. (It also preceded any consensus on its meaning and feasibility, but this is not a real problem, as "consensus" is never achieved on this Earth.)

Fisheries scientists and marine ecologists thus had to scramble to give the term some operational definition (see, e.g., Pikitch et al. 2004). Most settled on a list of essentials, including these: a place (for the ecosystem to be in) that is zoned (for different uses), whose integrity (especially with respects to sea floor communities) is to be protected from, e.g., trawling (at least in part; we are realists) and whose key forage species (e.g., krill or small fishes) are to be shared between humans and other predators, such as marine mammal and seabirds. Also, social scientists have joined in the fray, informing us of such things as "people are part of the ecosystem." (The quotation marks are to hint that this seemingly obvious statement is anything but, given that it prevents us from even conceiving of ecosystems that are deliberately left alone. Perhaps this is the intention?)

Out of this scramble came the realization that once an ecosystem becomes the focus of concern, there is no reason to privilege the fisheries, and thus ecosystem-based management (EBM) was born. However, this term hides more than it clarifies. Here is how a major environmental NGO defines "the principles of EBM" in one of its brochures: "Ecosystem-based management has objectives and targets that:

- Focus on maintaining the natural structure and function of ecosystems and their productivity;
- Incorporate human use and values of ecosystems in managing the resource;
- Recognize that ecosystems are dynamic and constantly changing;
- Are based on a shared vision of all stakeholders;
- Are based on scientific knowledge, adapted by continual learning and monitoring."

In other words, EBM as defined here means everything and nothing, i.e., the term has devolved toward vacuity. This is similar to *sustainability*, which, as originally proposed, could be taken to mean things being done such that they could remain more or less the way they are forever (or at least for a long time; see Chapter 4). Sustainability, however, has devolved to "sustainable growth," which is an oxymoron because something that grows (e.g., an economy, or fisheries catches) cannot continue to do so forever, or even for a long time. This is well illustrated by the unraveling, at the time of this writing, of the various Ponzi schemes of which Wall Street was so fond. The point about these various concepts is that their theoretical elaboration often substitutes for action to protect exploited stocks, or biodiversity in general, and the ecosystems in which they are embedded.

23. Why "allegedly"? The fluctuations of the Peruvian anchoveta really make interpretation of any trend data for the rest of the world difficult. Besides, our key graph (Figure 3.1A) did show trends with and without anchoveta.

24. While previous FAO analyses showed that "the large, high-value species landings decreased," we showed that the *total* catch was also declining. This was not old news, as alleged, but real news.

25. Let's note in passing that one major reason Africa cannot better manage their resource is because the exclusive economic zones of African countries are crawling with European and East Asian distant-water fleets, some paying nominal access fees (Kaczynski and Fluharty 2002), some not.

26. And this is a policy change that we achieved.

27. These private communications were actually hate mail, such as I occasionally get from trawl operators (especially from New England) after giving another interview about

the devastating effects of the gear on marine ecosystems. In this case, most of the mail dealt with the particularities of Chinese fisheries that we supposedly did not consider, such as the fact that invertebrates are targeted that are ignored in the West, or that the Chinese fisheries do not discard anything—in other words: the official line. The saddest of these private communications was from the director of a biochemistry laboratory, who wrote that the reason Chinese fishermen have "big catches" is because they work hard, as opposed to Africans who don't, etc. Clearly, we still have a long way to go.

28. Some great examples are available from Canada. J.A. Hutchings, C.J. Walters, and R.A. Haedrich, three university-based fisheries scientists, gave detailed background in Canada's leading fisheries journal of two cases (one of these being the collapse of northern cod) where bureaucrats of Canada's federal agency for researching and managing fisheries, the Department of Fisheries and Oceans (DFO), ignored, downplayed, and even suppressed scientific information that appeared to go against their industry-friendly policies (Hutchings et al. 1997). And to add insult to injury, an editorial in the issue of the journal that published the paper refers to an unsuccessful attempt to suppress that paper, as well.

29. The scientist/journalist pair here comprises Philippe Cury, a senior researcher at the French *Institut de Recherche pour le Développement* (IRD, which means what you think it does), and Yves Miserey, who has the "science beat" at *Le Figaro*, a major French daily. They cooperated on a book about the parlous state of the world fisheries, with emphasis on France (Cury and Miserey 2008). Among other things, it inspired much of an influential report by a French senator, aiming at a radical reform of French fisheries policy (Cléach 2008).

30. I have an Obama-like background (French mother, absent African American father, lived on different continents, etc.), with Dickensian elements added, which makes for halfway interesting profiles without my having to contribute anything really outrageous (Malakoff 2002; Yoon 2003).

Chapter 4

1. This obviously was a case of technical expertise in a narrow subfield being translated into an assumed ability to review one's entire discipline. The next step of rampant grandiosity is the notion that one can write about metaphysical questions, or even ... about the future (see Chapter 5).

2. Contribution originally authored by D. Pauly, V. Christensen, S. Guénette, T.J. Pitcher, U.R. Sumaila, C.J. Walters, R. Watson, and D. Zeller; published August 8, 2002; reprinted with permission.

3. In the original contribution, this was in the *early* 19th century, which was wrong. Only one person noticed and told me (thank you Peter Tyedmers).

4. I never heard anybody comment about this enormous number, or ask how it was derived. The story may be told here: Jaime Mendo is a Peruvian friend and colleague (and former PhD student) who mentioned in 1984 or so that he had a school friend who had worked over 10 years in various fish meal plants in Peru and who knew lots of workers in those plants. Thus, we devised a form where each step of the production of fish meal from Peruvian anchovy would be identified, and the difference between official numbers and the

actual values would be noted. This form was then filled in by the workers in the plants (no, we did not have permission—we would not have been given any—and yes, there was some politics involved). The results converged nicely (the central limit theorem again!) and were validated by Jaime, using fish meal export data. Thus, the annual catch of Peruvian anchovy was at least 18 million tonnes prior to the collapse, not 12 million tonnes as one finds in official accounts. Interestingly, this early "IUU" estimate of 50% [= 100 × 12/(18 − 12)] corresponds nicely to the estimates of IUU used to construct Figure 4.1.

5. Since the original publication of this graph, much work has been conducted at UBC's Fisheries Centre and elsewhere that will allow refining and validating this first global estimate of the catch of IUU fisheries. Thus, at the Fisheries Centre, the *Sea Around Us* Project has published "catch reconstructions" that take IUU fishery catches into account for numerous countries and territories (Zeller and Pauly 2007; Zeller et al. 2007), while the group led by T.J. Pitcher collaborated with the London-based MRAG in assembling a compendium of unreported catches from 59 countries and the high seas (Pramod et al. 2008) and estimating illegal catches (Agnew et al. 2009). Overall, these results, while representing work in progress, suggest that Figure 4.1 still gives a good approximation of the total catch of marine fisheries in the world.

6. Size spectrum models of ecosystems require field data, which are often difficult to assemble, at least if the spectrum in question is supposed to cover a wide range of sizes. Two approaches exist, however, to extract size spectra from Ecopath models of ecosystems, of which hundreds, covering a wide range of marine and freshwater ecosystems, already exist (www.ecopath.org). The first requires only that the asymptotic weight and the parameter K of the von Bertalanffy growth equation be known for every state variable (or "box") and be added to the model's input (Pauly and Christensen 2002). The output is a bivariate plot of biomass (all groups combined) vs. weight (in log units), whose slope will reflect the exploitation status of the ecosystem as a whole. The other approach is to reexpress an Ecopath model as a biomass (or biomass flow) vs. trophic level (Gascuel et al. 2009), which can then be used to explore ecosystem responses under various exploitation scenarios. The latter is still very much a work in progress, but it seems very promising.

7. This is also the reason why "culling" marine mammals, especially baleen whales, does not necessarily increase fisheries yields (Gerber et al. 2009). It is in fact bizarre that Japan has convinced many national delegations to the International Whaling Commission to support their "whales eat our fish" argument, but they have. As a result, what little research is conducted by the fishery departments of several Caribbean, West African, and South Pacific countries has been distorted into trying to blame whales for the depredations of local and distant-water fleets, while the FAO, of which Japan is an influential member, has had to organize several international meetings dealing with this useless nonissue.

8. Citing Casey and Myers (1998), who reported on the "[n]ear extinction of a large, widely distributed fish," the barndoor skate, previously abundant in deeper waters off the Canadian east coast and New England, may seem to have been impolitic, as its biomass, off New England, has recovered slightly since the mid-1990s (Boelke et al. 2005), i.e., when the time series ended upon which Casey and Myers (1998) based their analysis. This was used by some to contest their results, as if they could time travel. Actually, the belated increase

of this skate off New England is not surprising, since "[r]eductions in fishing effort, begin-ning in the mid 1990s in the New England area, were followed by increases in biomass of several groundfish and flounder stocks" (NOAA 2006). Still, we should have cited, instead, the eradication of all large skate from the Irish Sea (Brander 1981), which nobody contests, and whose demonstration was one of the first papers of this kind. Anyway, the point is now moot, as a clear pattern is established of the continued existence of large skate and rays sim-ply being incompatible with trawling (Cheung et al. 2005; Dulvy and Reynolds 2002).

9. Ecosystems can be "controlled" from the "top down" (i.e., their variability is caused by predators or a fishery exploiting them) or from the "bottom up" (i.e., their variability is caused by changes of primary production) or both ("mixed control"; Cury et al. 2008). Modeling tools now exist for evaluating the position of any aquatic ecosystem along the top-to-bottom control spectrum (Christensen and Walters 2004a), and hence there is no further justification for making a priori assumptions about the type of control occurring in an ecosystem. This is important since bottom-up control, which is the default assumption of many biological oceanographers (e.g., those who worked in GLOBEC), is also a risky assumption, as it downplays the potentially disrupting role of fisheries (which usually work in a top-down manner).

10. Shortly after the original version of this contribution was published, we realized the need to reestimate the amounts of subsidies made available to fisheries, as we considered the estimate used previously to be unreliable (notably because the estimators failed to include several kinds of subsidies, and most developing countries). This led to a global estimate of US$30–34 billion of government subsidies to fisheries per year for the period around 2003, of which US$20 billion are considered capacity-enhancing or "bad" subsidies (Sumaila et al. 2007).

11. Additional evidence for this was presented by Newton et al. (2007), who compared coral reef catch data from 49 tropical island countries of the world and concluded that "the Earth would require an additional 75,031 km^2 of coral reef area with the same productivity and resilience as the studied reefs to ensure that current catches are sustainable—an area that is equivalent to 3.7 Great Barrier Reefs." And their conclusions would have been even stronger if they had considered that coral reef fishery catches are grossly underreported by most countries (see, e.g., Zeller et al. 2007).

12. This was erroneously stated as "0.01%" in the original version of this contribution. The correct figure of 0.1% assumes that only about 10–20% of the nominally protected area, which amounts to 0.7% of the world (Wood et al. 2008), is effectively protected. Note that these numbers are not markedly affected by the declaration, in January 2008 by the outgoing US president, of large new MPAs around sparsely inhabited or uninhabited islands in the Pacific.

Chapter 5

1. Contribution originally authored by D. Pauly, J. Alder, E. Bennett, V. Christensen, P. Tyedmers, and R. Watson; published November 21, 2003. Reprinted with permission.

2. In fact, there is a growing movement in Peru to turn the local "anchoveta" from a mass

fish, fit only for reduction to fish meal and oil, into products for direct human consumption, ranging from canned fillets to fish paste, or marketed fresh for grilling, as is traditionally done, e.g., in Spain. There were efforts already in the early 1960s to turn anchoveta into human food, notably in United Nations' projects (Contesso 1965). They focused on powdered fish protein concentrate (FPC) that, so the idea went, would be added to bread and similar (subsidized) staples. These schemes never worked, and they certainly did nothing to change the poor public image of anchoveta, which in Peru was shaped by its use in institutional canteens, such as those in prisons.

My friend Patricia Majluf, a Peruvian professor and prize-winning conservationist, conceived of an alternative strategy, i.e., showing that anchoveta could be part of *tasty* meals. She convinced 20 leading chefs in Lima to invent new dishes featuring anchoveta, and to serve them in their restaurants. (See Pauly 2006a for an account of such dishes, worthy of *Babette's Feast*.) And she convinced the president of Peru to share with various dignitaries a meal of anchoveta, which was shown on national TV. This worked, a presidential decree was issued mandating that a fraction of the anchoveta catch be supplied fresh to markets, thus initiating the realizations of Patricia Majluf's (and Hall's 2007) vision. Now, freshly caught anchoveta are available in the markets of Lima, human consumption of anchoveta is increasing, and serious investment is being made into canneries, with frozen exports being next. The potential benefits to Peru will grow in proportion to the fraction of the anchoveta catch that is used for human consumption. Global food security will likewise benefit, since presently about one-third of global landings are turned into fish meal (Alder et al. 2008).

3. The results of the Millennium Ecosystem Assessment were published by Island Press in 2005. Its 900+ page volume on *Current State and Trends* was devoted overwhelmingly to terrestrial eco- and production systems; only two chapters were devoted to the marine realm, i.e., "Coastal Ecosystems" (Agardy et al. 2005) and "Marine Fisheries Systems" (Pauly et al. 2005).

4. The potential "rent" lost through overfishing has been estimated at US$50 billion (see also note 18, Chapter 1). However, it must also be mentioned that the cost of fishing routinely estimated by economists evaluating fisheries already includes return to labor (wages) and capital (as "standard" rate of return on investments), and hence what are lost through overfishing are, not "regular" profits, but Wall Street–inspired "extra" profits (Bromley 2009).

5. This is exactly the point of Jacquet and Pauly (2007).

6. This is obviously a wildly unrealistic assumption, as we are now locked in a future that will include significant warming (especially at high latitudes; Sarmiento et al. 2004) and other deleterious changes for marine biodiversity and ecosystems (and hence fisheries), including increased stratification (and thus reduced oxygenation; Shaffer et al. 2009) and acidification (Doney et al. 2009). The *Sea Around Us* Project has begun to deal with these issues (see Cheung et al. 2008a, 2008b, 2009, 2010), becoming, in the process, the first fisheries research group to do so on a global basis, i.e., at the appropriate scale (these are, after all, planetary processes). However, these issues are not followed up on here, except to point out that they are making it unlikely that present global fisheries catches will be sustained, even if the management of the underlying fisheries were to improve.

7. The latest "State of World Fisheries and Aquaculture" (FAO 2009), which presents

catch data up to 2006, contributes nothing that would lead one to question this view. Indeed, the trend of Figure 3 (i.e., excluding China) in FAO (2009) continues to 2006 the downward catch trend presented in our Figure 3.1 (upper panel) in Pauly et al. (2003).

Epilogue

1. The numbers of citations received (to the end of 2008) by these five contributions differ depending on whether one relies on Thompson ISI Web of Science (WS), which uses a large set of peer-reviewed journals as sources of citations, or Google Scholar (GS), which relies on all documents available online (Pauly and Stergiou 2005). These citations ("cites") and related indices are as follows:

Contribution	Age (years)	WS cites	GS cites	WS/yr	Refs	INK (WS/Refs)	INK/yr
1 PPR	13.8	333	504	24	24	13.9	0.97
2 Fishing down	10.9	812	1131	74	30	26.1	2.40
3 China	7.1	126	240	18	20	6.0	0.87
4 Sustainability	6.4	421	630	66	100	4.2	0.65
5 Future	5.1	105	145	21	27	3.9	0.76

As might be seen, it is contribution 2 that dominates the lot. Indeed, Branch (2009) lists contribution 2 among what he calls the "books and papers cited most often by fisheries scientists," while contribution 4 is listed among works "destined to become classics."

This table also shows that the numbers of citations identified by the Web of Science is lower than those identified by Google Scholar, and hence the "equivalence" of these two citation tracking services (Pauly and Stergiou 2005) is not verified, at least not for this small sample. However, the Index of New Knowledge (INK), obtained by dividing the *citations* received by an article by the number of *references* it cites + 1 (Pauly and Stergiou 2008), confirms Branch (2009) for contribution 2; it also shows that contribution 4, although much cited, generated the least "new knowledge," which is what you would expect from a review.

2. And what, finally, of the film that inspired the title of this book? I recently reconnected with an old friend, Rainer Vowe, who has a family background as complicated as mine and with whom I got along fine, notably because he is color blind (literally). He now teaches film critique at the University of Bochum, Germany, and thus I told him about the book I was writing, and the film from which its title is taken. Here is his overnight answer, freely translated from German:

A man comes home. Could Homer be the author of *Five Easy Pieces*? The hero is not as cleverly devious as Ulysses; rather, he is a nonconformist. Jack Nicholson, or whatever his name is [Bobby Duplea], lived on Mount Olympus among the Gods of Music (he played in concerts, no less—like Condoleezza Rice). But he descended into the underworld of the blue collars, and became a worker on an oil field. He might again ascend the musical Olympus; but first, he has a few woman issues to resolve, as did our King of Ithaca.

Now to semantics. The fact that this film by Bob Rafelson has this specific title also relates to music: "easy pieces" are études, musical pieces for beginners and advanced music students, who use them to practice and thus improve their skills. In the late 19th century, these études became more challenging and gradually turned into "studies," which, as in the case of Chopin's, were meant to be played before a public of connoisseurs. Thus, while the words *easy pieces* remained the same, the status of the pieces, and the actual degree of difficulty involved in their mastery, oscillated between beginners' pieces and studies showcasing superior skills.

Later, the term *easy pieces* was applied to popular music as well, and it can be found as part of the title of compilations of country music songs, or jazz. The film obviously taps into this. By taking over these words, and transposing them into a new context, it also asserts itself as presenting études or studies of a character and his milieu.

When using the title of this film as title of a (text?) book, you [DP] point to a film that itself points at something else, i.e., the world of music, and to both learning ("études") and mastery ("studies"). I guess you saw both, and the fact that the readers can move from one to the other. I presume this was what you wanted, was it not?

Note that, in addition to five classical pieces [see entry on *Five Easy Pieces* in Wikipedia], the film also features five songs by country singer Tammy Wynette, including her major hit "Stand by Your Man," which may provide some support for the notion that this film deals with two parallel worlds. But it is possible to over-interpret things, as well. After all, Nicholson [2006] said, in a conversation with Peter Bogdanovich: "What is *Five Easy Pieces*? It's a guy who starts off in disguise as a hick oil worker, and in the middle of the picture turns out to be an intellectual concert pianist. What could be more vehicular than that?"

Perhaps you could suggest that whoever reads your book will be turned from worker to artist or from a simple angler to a maritime philosopher.

Now, this is an ending!

Acknowledgments

I thank my colleagues in the FishBase and *Sea Around Us* projects for their collaboration on the contributions reprinted here. Particularly, I thank Villy Christensen, with whom I closely collaborated on four of these five contributions, and for his support (including some prose) toward the preparation of this book. I also thank Sandra Pauly and Jennifer Jacquet for insightful comments on the draft; Ms. Aque Atanacio, Los Baños, Philippines, for doing or redoing all the graphs included here; my long-term collaborator, Maria "Deng" Palomares, for assembling and analyzing the citation data used for the penultimate endnote; and my old friend Rainer Vowe for an exegesis of *Five Easy Pieces* (the film), featured in the last endnote.

The coauthors of the five contributions included in this book, which could not have been written without their initial input, deserve special thanks. They are V. Christensen (Chapters 1, 2, 4, and 5), Reg Watson (3, 4, and 5), J. Alder (5), E. Bennett (5), J. Dalsgaard (2), R. Froese (2), S. Guénette (4), F. Torres Jr. (2), T.J. Pitcher (4), U.R. Sumaila (4), P. Tyedmers (5), C.J. Walters (4), and D. Zeller (4). Other colleagues who helped with these contributions and who have my sincere gratitude are R.E. Ulanowicz for suggesting an approach to aggregate trophic flows in ecosystem models by trophic levels; T. Platt and F. Chavez for advice on global primary productivity and the nearly 100 authors of Ecopath models used for inferences (Chapter 1); H. Valtysson for discussions and A. Laborte for programming (Chapter 2); A. Gelchu for the distribution ranges of commercial fishes; V. Christensen for an upwelling index, and Nancy Baron for the outreach and the description thereof (Chapter 3); Gary Russ and M.L. Deng Palomares for various suggestions that improved the text (Chapter 4); and Adrian Kitchingman and Wilf Swartz for assistance with the data behind two figures (Chapter 5).

Thanks are also due to the institutions that funded some of my work and/or the work of the coauthors of these five contributions: the Natural Sciences and Engineering Research Council of Canada, the Danish International Development Agency, and the European Commission (Directorate-General for Development).

However, this book would not have been possible without the support of the Pew Charitable Trusts, and especially of Dr. Joshua Reichert, the Director of the Pew

Environment Group. They took a measure of risk in supporting my proposal of 1998 for the establishment of the *Sea Around Us* Project. It worked.

I also thank Mr. Todd Baldwin, for spending a weekend in Vancouver to help me focus on my message and for his careful editing of what I thought were successive final drafts, and my other friends at Island Press for turning my files, again, into a neat little book. Finally, I wish to thank Philippe Cury and the French *Institut de Recherche pour le Developpement* (IRD) for inviting me to spend the first three months of 2009 at the *Centre de Recherche Halieutique* in Sète, France, where the first draft was completed.

References

Agardy, T., J. Alder, P. Dayton, S. Curran, A. Kitchingman, M. Wilson, A. Catenazzi, J. Restrepo, C. Birkeland, S. Blaber, S. Saifullah, G. Branch, D. Boersma, S. Nixon, P. Dugan, N. Davidson and C. Vörösmarty. 2005. Coastal ecosystems, pp. 513–549. *In:* R. Hassan, R. Scholes and N. Ash (eds.), *Millennium Ecosystem Assessment: Ecosystems and Human Well-being: Current States and Trends,* Vol. 1. Island Press, Washington, DC.

Agnew, D.J., J. Pearce, G. Pramod, T. Peatman, R. Watson, J. Beddington and T.J. Pitcher. 2009. Estimating the worldwide extent of illegal fishing. *PLoS ONE* 4(2): e4570.DOI: 10.1371/journal .pone.0004570.

Albaret, J.-J. and R. Laë. 2003. Impact of fishing on fish assemblages in tropical lagoons: the example of the Ebrié lagoon, West Africa. *Aquatic Living Resources* 16(1): 1–9.

Alcala, A.C. and G.R. Russ. 1990. A direct test of the effects of protective management on abundance and yield of tropical marine resources. *Journal du Conseil international pour l'Exploration de la Mer* 46: 40–47.

Alder, J., B. Campbell, V. Karpouzi, K. Kaschner and D. Pauly. 2008. Forage fish: from ecosystems to markets. *Annual Reviews in Environment and Resources* 33: 153–166 [+ 8 pages of figures].

Allen, R.R. 1971. Relation between production and biomass. *Journal of the Fisheries Research Board of Canada* 28: 1573–1581.

Alverson, D.L., M.H. Freeberg, S.A. Murawski and J.G. Pope. 1994. A global assessment of fisheries by-catch and discards. *FAO Fisheries Technical Paper* 339, 233 p.

Alverson, D.L., A.R. Longhurst and J.A. Gulland. 1970. How much food from the sea? *Science* 168: 503–505.

Alverson, D.L. and W.T. Pereyra. 1969. Demersal fish exploration in the Northeastern Pacific Ocean—an evaluation of exploratory fishing methods and analytical approaches to stock size and yield forecast. *Journal of the Fisheries Research Board of Canada* 26: 1185–2001.

Arancibia, H. and S. Neira. 2005. Long-term changes in the mean trophic level of Central Chile fishery landings. *Scientia Marina* 69(2): 295–300.

Asimov, I. 1977. *The Planet That Wasn't: A Mind-dazzling Excursion into the Realm of Myth, Science and Speculation.* Discus Books, New York, 237 p.

Atkinson, M.J. and R.W. Grigg. 1984. Model of a coral reef ecosystem. II. Gross and net primary production at French Frigate Shoals, Hawaii. *Coral Reefs* 3: 13–22.

Baisre, J.A. 2000. Chronicles of Cuban Marine Fisheries (1935–1995): Trend analysis and fisheries potential. *FAO Fisheries Technical Paper* 394, 26 p.

Bakun, A. 1990. Global climate change and intensification of coastal ocean upwelling. *Science* 247: 198–201.

Baron, N. 1998. The Straits of Georgia. *The Georgia Straight* (Vancouver), September 3–10: 15–21.

Baron, N. 1999a. Looking for bigger fish to fry: An American foundation has granted UBC biologist

Daniel Pauly $3 million to ascertain the effect of current fisheries practices on our oceans, and how to restore abundance. *The Vancouver Sun,* November 5, 1999.

Baron, N. 1999b. Sea-ing Around Us: the "Pew Project" is under way. *The Sea Around Us Newsletter* (1): 1–4.

Baumann, M. 1995. A comment on transfer efficiencies. *Fisheries Oceanography* 4(3): 264–266.

Beddington, J.R. 1995. Fisheries—the primary requirements. *Nature* 374(6519): 213–214.

Beddington, J.R. and J.G. Cooke. 1983. The potential yield of fish stocks. *FAO Fisheries Technical Paper* 242, 47 p.

Beddington, J.R. and R.M. May. 1982. The harvesting of interacting species in a natural ecosystem. *Scientific American* 247: 62–69.

Bennett, E.M., S.R. Carpenter, G.D. Peterson, G.S. Cumming, M. Zurek and P. Pingali. 2003. Why global scenarios need ecology. *Frontiers in Ecology and the Environment* 1(6): 322–329.

Beverton, R.J.H. and S.J. Holt. 1957. *On the Dynamics of Exploited Fish Populations.* Chapman and Hall, London. Facsimile reprint 1993, 533 p.

Beverton, R.J.H. and S.J. Holt. 1964. Tables of yield functions for fishing assessment. *FAO Fisheries Technical Paper* 38, 49 p.

Bhathal, B. and D. Pauly. 2008. "Fishing down marine food webs" and spatial expansion of coastal fisheries in India, 1950–2000. *Fisheries Research* 91: 26–34.

Boelke, D., T. Gedamke, K. Sosebee, A. Valliere and B. Vanpelt. 2005. *Final Skate Annual Review.* New England Fishery Management Council, Newburyport, Mass.

Bohnsack, J.A. 1990. The potential of marine fisheries reserves for reef management in the US Southern Atlantic. *NOAA Technical Memorandum* NMFS-SEFC–261, 40 p.

Botsford, L.W., J.C. Castilla and C.H. Peterson. 1997. The management of fisheries and marine ecosystems. *Science* 277: 509–515.

Boyd, R.T. 1990. Demographic History 1774–1874, pp. 135–148. *In:* W. Suttles (ed.), *Handbook of American Indians: Northwest Coast.* Smithsonian Institute, Washington, DC.

Branch, T. 2009. Books and Paper Cited Most Often by Fisheries Scientists. http://knol.google.com/k/trevor-a-branch/books-and-papers-cited-most-often-by/3oifsmbbrsl2q/1?domain=knol.google.com&locale=en#.

Brander, K. 1981. Disappearance of common skate *Raja batis* from the Irish Sea. *Nature* 290: 48–49.

Bray, K. 2000. *A global review of illegal, unreported and unregulated (IUU) fishing. Expert consultation on illegal, unreported and unregulated fishing.* FAO, IUU/2000/6, Rome, 53 p.

Bromley, D. 2009. Abdicating responsibility: the deceits of fisheries policy. *Fisheries* 34(6): 280–290.

Caddy, J.F., F. Carocci and S. Coppola. 1998a. Have peak fishery production levels been passed in continental shelf areas? Some perspectives arising from historical trends in production per shelf area. *Journal of Northwest Atlantic Fishery Science* 23: 191–219.

Caddy, J.F., J. Csirke, S.M. Garcia and R.J. Grainger. 1998b. How pervasive is "Fishing Down Marine Food Webs?" *Science* 282: 1383a.

Caddy, J.F. and L. Garibaldi. 2000. Apparent changes in the trophic composition of world marine harvests: the perspective from the FAO capture database. *Ocean & Coastal Management* 43(8–9): 615–655.

Caddy, J.F., R. Refk and T. Dochi. 1995. Productivity estimates for the Mediterranean—evidence of accelerating ecological change. *Ocean & Coastal Management* 26(1): 1–18.

Caddy, J.F. and P.G. Rodhouse. 1998. Cephalopod and groundfish landings: evidence for ecological change in global fisheries? *Reviews in Fish Biology and Fisheries* 8(4): 431–444.

Carr, M.H. and D.C. Reed. 1993. Conceptual issues relevant to marine harvest refuges: Examples from temperate reef fishes. *Canadian Journal of Fisheries and Aquatic Science* 50(9): 2019–2028.

Casey, J.M. and R.A. Myers. 1998. Near extinction of a large, widely distributed fish. *Science* 281(5377): 690–692.

Castillo, S. and J. Mendo. 1987. Estimation of unregistered Peruvian anchoveta (*Engraulis ringens*) in official catch statistics, 1951 to 1982, pp. 109–116. *In:* D. Pauly and I. Tsukayama (eds.), *The Peruvian Anchoveta and Its Upwelling Ecosystem: Three Decades of Change*. ICLARM Studies and Reviews 15, Manila, Philippines.

CBD. 2004. The 2020 biodiversity target: a framework for implementation, p. 351. (Annex I, decision VII/30), *Decisions from the Seventh Meeting of the Conference of the Parties of the Convention on Biological Diversity, Kuala Lumpur, 9–10 and 27 February 2004*. CBD Secretariat, Montreal.

Chang, J. 1991. *Wild Swans: Three Daughters of China*. Simon and Schuster, London, 538 p.

Chapman V.M. 1965. Potential resources of the oceans. Van Camp Sea Food Co., Port of Long Beach CA. 43p.

Cheung, W.L., T.J. Pitcher and D. Pauly. 2005. A fuzzy logic expert system to estimate intrinsic extinction vulnerabilities of marine fishes to fishing. *Biological Conservation* 124: 97–111.

Cheung, W.W.L., C. Close, V. Lam, R. Watson and D. Pauly. 2008a. Application of macroecological theory to predict effects of climate change on global fisheries potential. *Marine Ecology Progress Series* 365: 187–193.

Cheung, W.W.L., V.W.Y. Lam and D. Pauly (eds.) 2008b. *Modelling Present and Climate-shifted Distribution of Marine Fishes and Invertebrates*. Fisheries Centre Research Report 16(3), 72 p.

Cheung, W.W.L., V.W.Y. Lam, J.L. Sarmiento, K. Kearney, R. Watson and D. Pauly. 2009. Projecting global marine biodiversity impacts under climate change scenarios. *Fish and Fisheries* DOI: 10.1111/j.1467–2979.00315x.

Cheung, W.W.L., V.W.Y. Lam, J.L. Sarmiento, K. Kearney, R. Watson, D. Zeller and D. Pauly. 2010. Large-scale redistribution of maximum fisheries catch potential in the global ocean under climate change. *Global Change Biology.* DOI: 10.1111/j.1365–2486.2009.01995.x.

Chen, W. 1999. Marine resources, their status of exploitation and management in the People's Republic of China. *FAO Fisheries Circular* 950, 60 p.

Choat, J.H. and D.R. Robertson. 2002. Age based studies, pp. 57–80. *In:* P.F. Sale (ed.), *Coral Reef Fishes: Dynamics and Diversity in a Complex Ecosystem*. Academic Press, San Diego.

Christensen, V. 1995a. Ecosystem maturity—towards quantification. *Ecological Modelling* 77(1): 3–32.

Christensen, V. 1995b. A model of trophic interactions in the North Sea in 1981, the Year of the Stomach. *Dana* 11(1): 1–28.

Christensen, V. 1996. Managing fisheries involving predator and prey species. *Reviews in Fish Biology and Fisheries* 6(4): 417–442.

Christensen, V. 1998. Fishery-induced changes in a marine ecosystem: insight from models of the Gulf of Thailand. *Journal of Fish Biology* 53: 128–142.

Christensen, V., S. Guénette, J.J. Heymans, C.J. Walters, R. Watson, D. Zeller and D. Pauly. 2003. Hundred-year decline of North Atlantic predatory fishes. *Fish and Fisheries* 4(1): 1–24.

Christensen, V. and J. Maclean. (eds.). 2004. Placing fisheries in their ecosystem context. *Ecological Modelling* 172(2–4; Special issue): 102–438.

Christensen, V. and D. Pauly. 1992a. Ecopath II—a software for balancing steady-state ecosystem models and calculating network characteristics. *Ecological Modelling* 61(3–4): 169–185.

Christensen, V. and D. Pauly. 1992b. A guide to the ECOPATH II program (version 2.1). *ICLARM Software* 6, 72 p.

Christensen, V. and D. Pauly (eds.). 1993a. *Trophic Models of Aquatic Ecosystems*. ICLARM Conference Proceedings 26, Manila, 390 p.

Christensen, V. and D. Pauly. 1993b. Flow characteristics of aquatic ecosystems, pp. 338–352. *In:*

V. Christensen and D. Pauly (eds.), *Trophic Models of Aquatic Ecosystems*. ICLARM Conference Proceedings 26, Manila.

Christensen, V., A. Trinidad-Cruz, J. Paw, J. Torres Jr. and D. Pauly. 1991. Catch and potential of major fisheries resource systems in tropical and subtropical areas. Appendix 6, pp. 13–43. *In: A Strategic Plan for International Fisheries Research*. International Center for Living Aquatic Resources Management, Manila.

Christensen, V. and C.J. Walters. 2004a. Ecopath with Ecosim: methods, capabilities and limitations. *Ecological Modelling* 172(2–4): 109–139.

Christensen, V. and C. J. Walters. 2004b. Trade-offs in ecosystem-scale optimization of fisheries management policies. *Bulletin of Marine Science* 74(3): 549–562.

Chuenpagdee, R., D. Preikshot, L. Liguori and D. Pauly. 2006. A public sentiment index for ecosystem management. *Ecosystems* 9: 463–473.

Clark, C.W. 1990. *Mathematical Bioeconomics: The Optimal Management of Renewable Resources*. Wiley, New York, 386 p.

Clark, W.G., S.R. Hare, A.M. Parma, P.J. Sullivan and R.J. Trumble. 1999. Decadal changes in growth and recruitment of Pacific halibut (*Hippoglossus stenolepis*). *Canadian Journal of Fisheries and Aquatic Sciences* 56(2): 242–252.

Cléach, M.-P. 2008. *Marée amère: pour une gestion durable de la pêche*. Office parlementaire d'évaluation des choix scientifiques et technologiques. No 1322 Assemblée nationale/No 132 Sénat, Paris, 175 p.

Cochrane, K.L. 2000. Reconciling sustainability, economic efficiency and equity in fisheries: the one that got away? *Fish and Fisheries* 1(1): 3–21.

Coelho, M.L. 2000. "Pandora's Box" in fisheries: is there a link between economy and ecology? pp. 33–35. *In:* F. Briand (ed.), *Fishing Down the Mediterranean Food Webs?* CIESM Workshop Series 12.

Coll, M., S. Libralato, S. Tudela, I. Palomera and F. Pronovi. 2008. Ecosystem Overfishing in the Ocean. *PLoS ONE* 3(12): e3881.

Contesso, G. 1965. Development of Fish Protein Concentrates in Peru. WHO/FAO/UNICEF Meeting, Rome, PAG, Doc 8/Add 18.

Costello, C., S.D. Gaines and J. Lynham. 2008. Can catch shares prevent fisheries collapses? *Science* 321: 1678–1681.

Cousins, S. 1985. Ecologists build pyramids again. *New Scientist* 4: 50–54.

Crossland, C.J., B.G. Hatcher and S.V. Smith. 1991. Role of coral reefs in global ocean production. *Coral Reefs* 10: 55–64.

Cury, P. and P. Cayré. 2001. Hunting became a secondary activity 2000 years ago: marine fishing did the same in 2021. *Fish and Fisheries* 2(2): 162–169.

Cury, P. and Y. Miserey. 2008. *Une mer sans poissons*. Calman-Lévy, Paris, 283 p.

Cury, P., L.J. Shannon and Y.-J. Shin. 2003. The functioning of marine ecosystems: a fisheries perspective, pp. 103–123. *In:* M. Sinclair and G. Valdimarsson (eds.), *Responsible Fisheries in the Marine Ecosystem*. CAB International, Wallingford.

Cury, P., Y.-J. Shin, B. Planque, J.M. Durant, J.M. Fromentin, S. Kramer-Schadt, N.-C. Stenseth, M. Travers and V. Grimm. 2008. Ecosystem oceanography for global change in fisheries. *Trends in Ecology and Evolution* 23(6): 338–346.

Cushing, D.H. 1987. *The Provident Sea*. Cambridge University Press, Cambridge, 329 p.

Dalzell, P. and D. Pauly. 1990. Assessment of the fish resources of Southeast-Asia, with emphasis on the Banda and Arafura Seas. *Netherlands Journal of Sea Research* 25(4): 641–650.

Daskalov, G.M. 2002. Overfishing drives a trophic cascade in the Black Sea. *Marine Ecology Progress Series* 225: 53–63.

Dawkins, R. 1996. *Climbing Mount Improbable*. Norton, New York, 312 p.

de Groot, S.J. 1971. On the interrelationships between morphology of the alimentary tract, food and feeding behavior in flatfishes (Pisces: Pleuronectiformes). *Netherlands Journal of Sea Research* 5(2): 121–196.

Dickie, L.M. 1972. Food chains and fish production, pp. 201–221. *In: Symposium on Environmental Conditions in the Northwest Atlantic, 1960–1969*. ICNAF Special Publications No. 8.

Dizon, L. and M.S.M. Sadorra. 1995. Pattern of publication by the staff of an international research center. *Scientometrics* 32(1): 67–75.

Doney, S.C., V.J. Fabry, R.A. Feely and J.A. Kleypas. 2009. Ocean Acidification: The Other CO_2 Problem. *Annual Review of Marine Science* 1: 169–192.

Doyle, A.C. 1902. *The Hound of the Baskervilles*. George Newnes, London, 243 p.

Du, S. 2001. *Nature* report unfounded: China Did Not Over-Report Fisheries Catch Statistics. Xinhua News Agency, Dec. 18, 2001.

Dugan, J.E. and G. E. Davis. 1993. Applications of marine refugia to coastal fisheries management. *Canadian Journal of Fisheries and Aquatic Sciences* 50(9): 2029–2042.

Dulvy, N.K., E. Chassot, S.J.J. Heymans, K. Hyde, E. Chassot, T. Platt and K. Sherman. 2009. Climate Change, Ecosystem Variability and Fisheries Productivity, Chapter 2, pp. 11–28. *In:* T. Platt, M.-H. Forget and V. Stuart (eds.), *Remote Sensing in Fisheries and Aquaculture: The Societal Benefits*. International Ocean-Colour Coordinating Group, Report No. 8, Dartmouth, Canada.

Dulvy, N.K. and J.D. Reynolds. 2002. Predicting extinction vulnerability in skates. *Conservation Biology* 16: 440–450.

Eckholm, E. 2001. Study says bad data inflated fishing yields. *The New York Times*, November 30, 2001.

Edeson, W.R. 1996. Current legal development: The Code of Conduct for Responsible Fisheries: an introduction. *International Journal of Marine and Coastal Law* 11: 233–238.

Eschmeyer, W.N. (ed.). 1998. Catalog of fishes. Special Publication, California Academy of Sciences, San Francisco. 3 vols. 2905 p. [available online, with regular updates at http://research.calacademy .org/ichthyology/catalog]

Essington, T.E., A.H. Beaudreau and J. Wiedemann. 2006. Fishing through marine food webs. *Proceedings of the National Academy of Sciences* 103: 3171–3175.

FAO. 1984. *Yearbook of Fishery Statistics. Catches and Landings [1982]*. Food and Agriculture Organization of the United Nations, Rome. Vol. 54, 393 p.

FAO. 1992. Review of the State of World Fisheries Resources, Part 1: the Marine Resources. *FAO Fisheries Circular* No. 710, rev. 8.

FAO. 1993a. Improving the fisheries contributions to world food supplies. FAO Fisheries Bulletin 6(5): 159–192.

FAO. 1993b. *Yearbook of Fishery Statistics. Catches and Landings [1991]*. Food and Agriculture Organization of the United Nations, Rome. Vol. 72, 653 p.

FAO. 1994. World review of highly migratory species and straddling stocks. *FAO Fisheries Technical Paper* 337, 74 p.

FAO. 1995. Code of Conduct for Responsible Fisheries. FAO, Rome, 41 p. http://www.fao.org/fi/ agreem/codecond/codecon.asp.

FAO. 1996. *Yearbook of Fishery Statistics. Catches and Landings [1994]*. Food and Agriculture Organization of the United Nations, Rome. Vol. 78, 357 p.

FAO. 1997a. Review of the state of world fishery resources: Marine fisheries. *FAO Fisheries Circular* 920, 173 p.

FAO. 1997b. *Yearbook of Fishery Statistics. Catches and Landings [1995].* Food and Agriculture Organization of the United Nations, Rome. Vol. 80, 183 p.

FAO. 1999. *FAO's fisheries agreements register (FARISIS).* Committee on Fisheries, 23rd Session, 15–19 February 1999 (COFI/99/Inf. 9E), FAO, Rome, 4 p.

FAO. 2000a. Fishery trade flow (1995–1997) for selected countries and products. *FAO Fisheries Circular* C961, Rome, 330 p.

FAO. 2000b. FISHSTAT Plus. Universal software for Fishery statistical time series. Version 2.3. Food and Agriculture Organization of the United Nations, Fisheries Department, Fishery Information, Data and Statistics Unit, Rome.

FAO. 2000c. Atlas of tuna and billfishes. http://www.fao.org/fishery/statistics/tuna-atlas/3/en.

FAO. 2001a. *State of the World Fisheries and Aquaculture 2000,* FAO, Rome, 142 p.

FAO. 2001b. Proceedings of the National Seminar on the System of Food and Agriculture Statistics in China, Beijing, 23–24 September 1999. Volumes I and II. Field Document No. 2/CHN/3 of Project GCP/RAS/171/JPN, Improvement of Agricultural Statistics in Asia and Pacific Countries. FAO, Rome.

FAO. 2002. Fishery statistics: reliability and policy implications. Available online at http://www.fao.org/DOCREP/FIELD/006/Y3354M/Y3354M00.HTM.

FAO. 2009. The State of World Fisheries and Aquaculture 2008. FAO, Rome, 84 p.

Feigon, L. 2000. A harbinger of the problems confronting China's economy and environment: the great Chinese shrimp disaster of 1993. *Journal of Contemporary China* 9(24): 323–332.

Fonteneau, A. 1997. *Atlas des pêcheries thonières tropicales: captures mondiales et environnement / Atlas of tropical tuna fisheries: world catches and environment.* ORSTOM, Paris, 192 p.

Forrest, R., T. Pitcher, R. Watson, H. Valtysson and S. Guénette. 2001. Estimating illegal and unreported catches from marine ecosystems: two case studies, pp. 81–93. *In:* T. Pitcher, U.R. Sumaila and D. Pauly (eds.), *Fisheries Impacts on North Atlantic Ecosystems: Evaluations and Policy Exploration.* Fisheries Centre Research Report 9(5).

Fox, M. 1995. *Oceans need protected areas to recover—study.* Reuters News.

Francis, R.C. 1974. Relationship of fishing mortality to natural mortality at the level of maximum sustainable yield under the logistic stock production model. *Journal of the Fisheries Research Board of Canada* 31: 1539–1542.

Frazier, J.G. 1997. Sustainable development: modern elixir or sack dress? *Environmental Conservation* 24(2): 182–193.

Froese, R. and D. Pauly (eds.). 1997. *FishBase 97: Concepts, design and data sources.* ICLARM, Manila, Philippines, 257 p.

Froese, R. and D. Pauly (eds.). 1998. *FishBase 98: Concepts, design and data sources.* ICLARM, Manila, 293 p.

Froese, R. and D. Pauly (eds.). 1999. *FishBase 99: Concepts, structure, et sources de données.* Traduit par N. Bailly et M.L.D. Palomares. ICLARM, Manille, 324 p.

Froese, R. and D. Pauly (eds.). 2000. *FishBase 2000: Concepts, design and data sources.* ICLARM, Los Baños, Philippines, 346 p.

Froese, R. and D. Pauly (eds.). 2006. *FishBase. World Wide Web electronic publication;* www.fishbase.org.

Fry, B. 1988. Food web structure on Georges Bank from stable C, N and S isotopic compositions. *Limnology & Oceanography* 33: 1182–1189.

Garcia, S.M. and C. Newton. 1997. Current situation, trends and prospects in world capture fisheries, pp. 3–27. *In:* E.L. Pikitch, D.D. Huppert and M.P. Sissenwine (eds.), *Global Trends: Fisheries Management.* American Fisheries Society Symposium, 20, Bethesda, Maryland.

Garfield, E. 1975. The obliteration phenomenon in science—and the advantage of being obliterated! *Current Contents* (51/52): 5–7. [Reprinted in Vol. 2, p. 396–398 of E. Garfield, *Essays of an Information Scientist*, ISI Press, Philadelphia.]

Gascuel, D., L. Tremblay-Boyer and D. Pauly. 2009. *EcoTroph (ET): a trophic-level based software for assessing impacts of fishing on aquatic ecosystems.* Fisheries Centre Research Report 17(1), 82 p.

Gee, H. 2002. Food and the future. *Nature* 418: 667.

Gerber, L.R., L. Morissette, K. Kaschner and D. Pauly. 2009. Should whales be culled to increase fishery yield? *Science* 323(5916): 880–881.

Gislason, H., M. Sinclair, K. Sainsbury and R. O'Boyle. 2000. Symposium overview: incorporating ecosystem objectives within fisheries management. *ICES Journal of Marine Science* 57(3): 468–475.

Gjøsaeter, J. and K. Kawaguchi. 1980. A review of the world resources of mesopelagic fish. *FAO Fisheries Technical Paper* 193, 151 p.

Gladwell, M. 2008. *Outliers: the Story of Success.* Little, Brown and Company, New York, ix, 309 p.

Golley, F.B. 1993. A history of the ecosystem concept in ecology: more than the sum of its parts. Yale University Press, New Haven, 254 p.

Gordon, H.S. 1954. The economic theory of a common property resource: the fishery. *Journal of Political Economy* 62: 124–142.

Gordon, R.A., M. Bernhard, M.J. Crawley, Y. Halim, K.J. Hsü, W.R. Jordan III, P. Kafka, H. Nöthel, D. Pauly, S.L. Pimm, G.S. Sayler and W. Van den Daele. 1991. Does bioscience threaten ecological integrity? pp. 185–201. *In:* D.J. Roy, B.E. Wynne and R.W. Old (eds.), *Bioscience ⇌ Society.* John Wiley & Sons, Chichester.

Graham, H.W. and R.L. Edwards. 1962. The world biomass of marine fishes, pp. 3–8 *In:* E. Heen and R. Kreuzer (eds.), *Fish in Nutrition.* Fishing News (Books), London.

Grainger, R.J.R. and S.M. Garcia. 1996. Chronicles of marine fishery landings (1950–1994). Trend analysis and fisheries potential. *FAO Fisheries Technical Paper* 359, 51 p.

Guénette, S., T.J. Pitcher and C.J. Walters. 2000. The potential of marine reserves for the management of northern cod in Newfoundland. *Bulletin of Marine Science* 66(3): 831–852.

Gulland, J.A. 1970. Summary, pp. 307–319. *In:* J.A. Gulland (ed.), The fish resources of the ocean. *FAO Fisheries Technical Paper* 97.

Gulland, J.A. (ed.). 1971. *The Fish Resources of the Oceans.* Fishing News Books, West Byfleet, UK.

Hall, M. 2007. 10 solutions to save the oceans: eat more anchovies. *Conservation* 8(3): 24.

Hall, S.J. 1998. *The Effects of Fisheries on Ecosystems and Communities.* Blackwell, Oxford. 274 p.

Hanneson, R. 2005. Right-based fishing: use rights versus property rights to fish. *Reviews in Fish Biology and Fisheries* 15(3): 231–241.

Hansson, S. and L.G. Rudstam. 1990. Eutrophication and Baltic fish communities. *AMBIO, a Journal of the Human Environment* 19(3): 123–125.

Hardin, D. 1968. The Tragedy of the Commons. *Science* 162: 1243–1248.

Hardy, A. 1956. *The Open Sea.* Collins, London, 335 p.

Heath, M.R. 2005. Changes in the structure and function of the North Sea fish food web, 1973–2000, and the impacts of fishing and climate. *ICES Journal of Marine Science* 62(5): 847–868.

Heinberg, R. 2003. *The Party's Over: Oil, War and the Fate of Industrial Societies.* New Society, Gabriola Island, BC, Canada. xii, 275 p.

Henderson, D.R., R.M. McNab and T. Rózsá. 2005. The hidden inequality in socialism. *The Independent Review* 9(3): 329–412.

Hites, R.A., J.A. Foran, D.O. Carpenter, M.C. Hamilton, B.A. Knuth and S.J. Schwager. 2004. Global assessment of organic contaminants in farmed salmon. *Science* 303: 226–229.

Hoar, W.S., D.J. Randall and J.R. Brett (eds.). 1979. *Fish Physiology*, Vol. VIII. Bioenergetics and Growth. Academic Press, New York, xvii+786 p.

Holt, S.J., J.A. Gulland, C. Taylor and S. Kurita. 1959. A standard terminology for fisheries dynamics. *Journal du Conseil International pour l'Exploration de la Mer* 24: 239–242.

Huang, B. and C.J. Walters. 1983. Cohort analysis and population dynamics of large yellow croaker in the China Sea. *North American Journal of Fisheries Management* 3: 295–305.

Hutchings, J.A. 2000. Collapse and recovery of marine fishes. *Nature* 406: 882–885.

Hutchings, J.A., C.J. Walters and R.L. Haedrich. 1997. Is scientific inquiry incompatible with government information control? *Canadian Journal of Fisheries and Aquatic Sciences* 54: 1198–1210.

Idyll, C.P. 1978. *The Sea Against Hunger*. Thomas U. Crowell, New York. 222 p.

Insightful Corporation, 2001. *S-Plus 6 for Windows Guide to Statistics*, Volume 1. Seattle, WA. xx+712 p.

Irwin, L.N. and D. Schulze-Makuch. 2003. Strategy for modeling putative multilevel ecosystem on Europa. *Astrobiology* 3(4): 813–821.

Iverson, R.L. 1990. Control of marine fish production. *Limnology and Oceanography* 35: 1593–1604.

Jackson, J.B.C., M.X. Kirby, W.H. Berger, K.A. Bjorndal, L.W. Botsford, B.J. Bourque, R.H. Bradbury, R. Cooke, J. Erlandson, J.A. Estes, T.P. Hughes, S. Kidwell, C.B. Lange, H.S. Lenihan, J.M. Pandolfi, C.H. Peterson, R.S. Steneck, M.J. Tegner and R.R. Warner. 2001. Historical overfishing and the recent collapse of coastal ecosystems. *Science* 293: 629–638.

Jacquet, J. and D. Pauly. 2007. The rise of seafood awareness campaigns in an era of collapsing fisheries. *Marine Policy* 31: 308–313.

Jacquet, J. and D. Pauly. 2008. Trade secrets: renaming and mislabeling of seafood. *Marine Policy* 32: 309–318.

Jarre, A., P. Muck and D. Pauly. 1991. Two approaches for modelling fish stock interactions in the Peruvian upwelling ecosystem. *ICES Marine Science Symposia* 193: 178–184.

Jarre-Teichmann, A. and V. Christensen. 1998. Comparative modelling of trophic flows in four large upwelling ecosystems: global versus local effects, pp. 423–443. *In:* M.H. Durand, P. Cury, R. Mendelssohn, C. Roy, A. Bakun and D. Pauly (eds.), *Global versus Global Change in Upwelling Areas*. Séries Colloques et Séminaires, ORSTOM Editions, Paris.

Jones, R. 1982. Ecosystems, food chains and fish yields, pp. 195–240. *In:* D. Pauly and G.I. Murphy (eds.), *Theory and Management of Tropical Fisheries*. ICLARM Conference Proceedings 9, Manila.

JRC. 2000. Joint Research Centre of the European Commission Space Applications Institute (Mar. Env. Unit). www.gmes.jrc/download/kyoto_prot/glob.marine.pdf.

Kaczynski, V.M. and D.L. Fluharty. 2002. European policies in West Africa: who benefits from fisheries agreements? *Marine Policy* 26: 75–93.

Karachle, P.K. and K.I. Stergiou. 2006. The effect of season and sex on trophic levels of marine fishes. *Journal of Fish Biology* 72(6): 1463–1487.

Kaschner, K. and D. Pauly. 2005. Competition between marine mammals and fisheries: food for thought, pp. 95–117. *In:* D.J. Salem and A.N. Rowan (eds.), *The State of Animals III: 2005*. Humane Society Press, Washington, DC.

Kelleher, K. 2004. Discards in the world's marine fisheries. An update. *FAO Fisheries Technical Paper* 470, 131 p.

Kennedy, D. 2006. Life on a human-dominated planet, pp. 5–12. *In:* D. Kennedy and the editors of Science Magazine (eds.), *Science Magazine's State of the Planet, 2006–2007*. AAAS and Island Press, Washington, DC.

Kennedy, D. and the editors of Science Magazine (eds.), *Science Magazine's State of the Planet, 2006–2007*. AAAS and Island Press, Washington, DC. xi+200 p.

Kesteven, G.K. and S.J. Holt. 1955. A note on the fisheries resources of the North West Atlantic. *FAO Fisheries Papers* 7, 11 p.

Kirkwood, G.P., J.R. Beddington and J.A. Rossouw. 1994. Harvesting species of different lifespans, pp. 199–227. *In:* P.J. Edwards, R.M. May and N.R. Webb (eds.), *Large Scale Ecology and Conservation Biology.* 35th Symposium of the British Ecological Society. Blackwell Scientific, Oxford.

Kline, T.C. Jr. and D. Pauly. 1998. Cross-validation of trophic level estimates from a mass-balance model of Prince William Sound using $^{15}N/^{14}N$ data, pp. 693–702. *In:* T.J. Quinn II, F. Funk, J. Heifetz, J.N. Ianelli, J.E. Powers, J.F. Schweigert, P.J. Sullivan and C.-I. Zhang (eds.), *Proceedings of the International Symposium on Fishery Stock Assessment Models.* Alaska Sea Grant College Program Report No. 98-01, Anchorage.

Kolding, J. 1993. Trophic interrelationships and community structure at two different periods of the Lake Turkana, Kenya: a comparison using the ECOPATH II box model, pp. 116–123. *In:* V. Christensen and D. Pauly (eds.), *Trophic Models of Aquatic Ecosystems.* ICLARM Conference Proceedings 26.

Koslow, J.A., G. Boehlert, J.D.M. Gordon, R.L. Haedrich, P. Lorance and N. Parin. 1999. Continental slope and deep-sea fisheries: implications for a fragile ecosystem. *ICES Journal of Marine Science* 57: 548–557.

Kurlansky, M. 2008. *The Last Fish Tale: The Fate of the Atlantic and Survival in Gloucester, America's Oldest Fishing Port and Most Original Town.* Random House, New York, 269 p.

Kwong, J. 1997. *The Political Economy of Corruption in China.* M.E. Sharpe, Armonk, New York. 175 p.

Larkin, P.A. 1977. An epitaph for the concept of maximum sustainable yield. *Transactions of the American Fisheries Society* 106(1): 1–11.

Laurans, M., D. Gascuel, E. Chassot and D. Thiam. 2004. Changes in the trophic structure of fish demersal communities in West Africa in the three last decades. *Aquatic Living Resources* 17(2): 163–173.

Law, R. 2000. Fishing, selection, and phenotypic evolution. *ICES Journal of Marine Science* 57(3): 659–668.

Lawton, J. 1994. What will you give up? *Oikos* 71: 353–354.

Lewis, J.B. 1977. Processes of organic production on coral reefs. *Biological Reviews of the Cambridge Philosophical Society* 52: 305–307.

Li, H. 2001. Does China Over-Report Fisheries Catch Statistics? *Southern Weekend* (Guangzhou). December 13, 2001.

Li, W.K.W., D.V. Subba Rao, W.G. Harrison, J.C. Smith, J.J. Cullen, B. Irwin and T. Platt. 1983. Autotrophic picoplankton in the tropical ocean. *Science* 219: 292–295.

Libralato, S., M. Coll, S. Tudela, I. Palomera and F. Pranovi. 2008. Novel index for quantification of fishing as removal of secondary production. *Marine Ecology Progress Series* 355: 107–129.

Libralato, S., F. Pranovi, S. Raicevich, F. Da Ponte, O. Giovanardi, R. Pastres, P. Torricelli and D. Mainardi. 2004. Ecological stages of the Venice Lagoon analysed using landing time series data. *Journal of Marine Systems* 51(1–4): 331–344.

Lieth, H. 1978. Primary productivity in ecosystems: comparative analysis of global patterns, pp. 300–321. *In:* H.F. Lieth (ed.), *Patterns of Primary Production in the Biosphere.* Dowden, Hutchinson & Ross, Stroudsburg, Pennsylvania.

Lindeman, R.I, 1942. The trophic-dynamic aspect of ecology. *Ecology* 23(4): 399–418.

Lindholm, J.B.I., P.J. Auster and L. Kaufman. 1999. Habitat-mediated survivorship of 0-year Atlantic cod (*Gadus morhua*). *Marine Ecology Progress Series* 180: 247–255.

Lipcius, R.N., R.D. Seitz, W.J. Goldsborough, M.M. Montane and W.T. Stockhausen. 2001. A deepwater dispersal corridor for adult female blue crabs in Chesapeake Bay, pp. 643–666. *In:*

D. Witherell (ed.), *Spatial Process and Management of Marine Populations.* Alaska Sea Grant College Program, AKK-SG-01-02, Anchorage.

Litzow, M. and D. Urban. 2009. Fishing through (and up) Alaskan food webs. *Canadian Journal of Fisheries and Aquatic Sciences* 66(2): 210–211.

Loder, N. 2001. The world's fish catch may be much smaller than previously thought. *The Economist* December: 1–7, 75–76.

Lomborg, B. 2001. *The Skeptical Environmentalist.* Cambridge University Press, Cambridge, 515 p.

Longhurst, A. 1998a. Cod: Perhaps if we all stood back a bit? *Fisheries Research* 38(2): 101–108.

Longhurst, A.R. 1998b. *Ecological Geography of the Sea.* Academic Press, San Diego, 398 p.

Longhurst, A.R. and D. Pauly. 1987. *Ecology of Tropical Oceans.* Academic Press, San Diego, 407 p.

Lotze, H.K. and I. Milewski. 2004. Two centuries of multiple human impacts and successive changes in a North Atlantic food web. *Ecological Applications* 14(5): 1428–1447.

Ludwig, D., R. Hilborn and C. Walters. 1993. Uncertainty, resource exploitation, and conservation: lessons from history. *Science* 260: 17–18.

Lyman, C.P., M.J. Gibbons, B.E. Axelsen, C.A.J. Sparks, J. Coetzee, B.G. Heywood and A.S. Brierley. 2006. Jellyfish overtake fish in a heavily fished ecosystem. *Current Biology* 16: R492–R493.

Mace, P.M. 1997. Developing and sustaining world fisheries resources: the state of fisheries and management, pp. 1–20. *In:* D.H. Hancock, D.C. Smith and J. Beumer (eds.), *Developing and Sustaining World Fisheries Resources.* Proceedings of the 2nd World Fisheries Congress, CSIRO Publishing, Collingwood, Australia.

Mace, P.M. 2001. A new role for MSY in single-species and ecosystem approaches to fisheries stock assessment and management. *Fish and Fisheries* 2(1): 2–32.

Macinko, S. and D.W. Bromley. 2002. *Who Owns America's Fisheries?* Island Press, Washington, DC, 48 p.

Maclean, J. (ed.). 1997. *Rice Almanac,* 2nd Edition, International Rice Research Institute, Los Baños, Philippines, 181 p.

Malakoff, D. 2002. Going to the edge to protect the sea. *Science* 296: 458–461.

Maranger, R., N. Caraco, J. Duhamel and M. Amyot. 2008. Nitrogen transfer from sea to land via commercial fisheries. *Nature Geoscience* 1: 111–112.

May, R.M. 1976. Patterns in multi-species communities, Chapter 8, pp. 142–162. *In:* R.H. May (ed.), *Theoretical Ecology: Principles and Applications.* Blackwell Scientific Publications, Oxford.

May, R.M. (ed.). 1984. *Exploitation of Marine Communities: Report of the Dahlem Workshop, Berlin, April 1–6, 1984.* Springer-Verlag, Berlin, x+367 p.

May, R.M., J.R. Beddington, C.W. Clark, S.J. Holt and R.M. Laws. 1979. Management of multispecies fisheries. *Science* 20: 267–277.

McCall-Smith, A. 1999. *The No. 1 Ladies' Detective Agency.* Polygon, Edinburgh, 256 p.

McClatchie, S. 1988. Food-limited growth of *Euphausia superba* in Admiralty Bay, South Shetland Islands, Antarctica. *Continental Shelf Research* 8(4): 329–345.

Mendoza, J.J. 1993. A preliminary biomass budget for the northeastern Venezuela shelf ecosystem, pp. 285–297. *In:* V. Christensen and D. Pauly (eds.), *Trophic Models of Aquatic Ecosystems.* ICLARM Conference Proceedings 26.

Merton, R. 1968. *Social Theory and Social Structure.* Free Press, New York, xxiii+702 p.

Mesek, G. 1962. Importance of fisheries production and utilization in the food economy, pp. 23–38. *In:* E. Heen and R. Kreuzer (eds.) fish *in* Nutrition, Fishing News Books, London.

Milazzo, M. 1998. Subsidies in world fisheries: a re-examination. *World Bank Technical Paper* 406, 86 p.

Milessi, A.C., H. Arancibia, S. Neira and O. Defeo. 2005. The mean trophic level of Uruguayan landings during the period 1990–2001. *Fisheries Research* 74(1–3): 223–231.

Mills, C.E. 2001. Jellyfish blooms: are populations increasing globally in response to changing ocean conditions? *Hydrobiologia* 451(1–3): 55–68.

Minagawa, M. and E. Wada. 1984. Stepwise enrichment of ^{15}N along food chains: further evidence and the relation between ^{15}N and age. *Geochimica and Cosmochimica Acta* 48: 1135–1140.

Moiseev, P.A. 1969. *The Living Resources of the World Ocean.* Pishchevaia promyshlannost, Moskva. 338 p. [Translated edition 1971 by Israel Program for Scientific Translation, Jerusalem, 334 p.]

Moiseev, P.A. 1994. Present fish productivity and bioproduction potential of the world aquatic habitats, pp. 70–75. *In:* C.W. Voigtlander (ed.), *The State of the World's Fisheries Resources.* Proceedings of the World Fisheries Congress. Plenary Sessions, Oxford and IBH, New Delhi.

Morgan, L. and R. Chuenpagdee. 2003. *Shifting Gears: Addressing the Collateral Impacts of Fishing Methods in U.S. Waters.* Island Press, Washington, DC, 42 p.

Morowitz, H.J. 1992. *The Thermodynamics of Pizza: Essays on Science and Everyday Life.* Rutgers University Press, New Brunswick, NJ, 247 p.

Mosquera, I., I.M. Coté, S. Jennings and J.D. Reynolds. 2000. Conservation benefits of marine reserves for fish populations. *Animal Conservation* 3: 321–332.

Muck, P. 1989. Major trends in the pelagic ecosystem off Peru and their implications for management, pp. 386–403. *In:* D. Pauly, P. Muck, J. Mendo and I. Tsukayama (eds.), *The Peruvian Upwelling Ecosystem: Dynamics and Interactions.* ICLARM Conference Proceedings 18, Manila.

Munro, G.R. 1979. The optimal management of transboundary renewable resources. *Canadian Journal of Economics* 12: 355–376.

Munro, G.R. and U.R. Sumaila. 2001. Subsidies and their potential impact on the management of the ecosystems of the North Atlantic, pp. 10–27. *In:* T.J. Pitcher, U.R. Sumaila and D. Pauly (eds.), *Fisheries Impacts on North Atlantic Ecosystems: Evaluations and Policy Exploration.* Fisheries Centre Research Report 9(5).

Murawski, S.A., R. Brown, H.L. Lai, P.R. Rago and L. Hendrickson. 2000. Large-scale closed areas as a fishery management tool in temperate marine systems: the Georges Bank experience. *Bulletin of Marine Science* 66: 775–798.

Myers, R.A., J.A. Hutchings and N.J. Barrowman. 1997. Why do fish stocks collapse? The example of cod in Atlantic Canada. *Ecological Applications* 7(1): 91–106.

Myers, R.A. and B. Worm. 2003. Rapid worldwide depletion of predatory fish communities. *Nature* 423: 280–283.

Naylor, R.L., R.J. Goldburg, J.H. Primavera, N. Kautsky, M.C.M. Beveridge, J. Clay, C. Folke, J. Lubchenco, H. Mooney and M. Troell. 2000. Effect of aquaculture on world fish supplies. *Nature* 405: 1017–1024.

Neutel, A.-M., J.A.P. Heesterbeek and P.C. de Ruiter. 2002. Stability in real food webs: weak links in long loops. *Science* 296: 1120–1123.

Newton, K., I.M. Cote, G.M. Pilling, S. Jennings and N.K. Dulvy. 2007. Current and future sustainability of island coral reef fisheries. *Current Biology* 17: 655–658.

Nicholson, J. 2006. I always have a plan, and it always goes better than my plan: Interview with Peter Bogdanovich. *Süddeutsche Zeitung Magazin* (46); http://sz-magazin.sueddeutsche.de/texte/anzeigen/1978.

NOAA. 2000. National Oceanographic and Atmospheric Agency of the USA, Marine Atlas; http://www.nodc.noaa.gov/OC5/data_woa.html.

NOAA. 2006. Status of Fishery Resources off the Northeastern US: Aggregate Resource and Landing Trends. www.nefsc.noaa.gov/sos/agtt/.

Odum, H.T. 1988. Self-organization, transformity and information. *Science* 242: 1132–1139.

Odum, W.E. and E.J. Heald. 1975. The detritus-based food web of an estuarine mangrove community, pp. 265–286, Vol. 1. *In:* L.E. Cronin (ed.), *Estuarine Research.* Academic Press, New York.

Okey, T.A. 2003. Membership of the eight Regional Fishery Management Councils in the United States: are special interests over-represented? *Marine Policy* 27(3): 193–206.

Orensanz, J.M.L., J. Armstrong, D. Armstrong and R. Hilborn. 1998. Crustacean resources are vulnerable to serial depletion—the multifaceted decline of crab and shrimp fisheries in the Greater Gulf of Alaska. *Reviews in Fish Biology and Fisheries* 8(2): 117–176.

Pace, M.L., J.J. Cole, S.R. Carpenter and J.F. Kitchell. 1999. Trophic cascades revealed in diverse ecosystems. *Trends in Ecology & Evolution* 14(12): 483–488.

Pang, L. and D. Pauly. 2001. Part 1 Chinese marine capture fisheries from 1950 to the late 1990s: the hopes, the plans and the data, pp. 1–27. *In:* R. Watson, L. Pang and D. Pauly (eds.), *The Marine Fisheries of China: Development and Reported Catches.* Fisheries Centre Research Report 9(2).

Parrish, R. 1989. The South Pacific Oceanic Horse Mackerel (*Trachurus picturatus murphyi*), pp. 321–331. *In:* D. Pauly, P. Muck, J. Mendo and I. Tsukayama (eds.), *The Peruvian Upwelling Ecosystem: Dynamics and Interactions.* ICLARM Conference Proceedings 18, Manila.

Parrish, R.H. 1998. Life history strategies for marine fishes in the late Holocene, pp. 525–535. *In:* M.H. Durand, P. Cury, R. Mendelssohn, C. Roy, A. Bakun and D. Pauly (eds.), *Local versus Global Changes in Upwelling Systems.* Séries Colloques et Séminaires. ORSTOM Editions, Paris.

Parsons, T.R. 1996. The impact of industrial fisheries on the trophic structure of marine ecosystems, pp. 352–357. *In:* G.A. Polis and K.D. Winemiller (eds.), *Food Webs: Integration of Patterns and Dynamics.* Chapman and Hall, New York.

Parsons, T.R. and Y.L.L. Chen. 1994. Estimates of trophic efficiency, based on the size distribution of phytoplankton and fish in different environments. *Zoological Studies* 33: 296–301.

Parsons, T.R., M. Takahashi and B. Hargrave. 1984. *Biological Oceanographic Processes.* Pergamon Press, Oxford, x+166 p.

Pauly, D. 1975. On the ecology of a small West African lagoon. *Berichte der Deutschen Wissenschaftlichen Kommission für Meeresforschung* 24(1): 46–62.

Pauly, D. 1979a. Theory and management of tropical multispecies stocks: a review, with emphasis on the Southeast Asian demersal fisheries. *ICLARM Studies and Reviews* 1, 35 p.

Pauly, D. 1979b. Biological overfishing of tropical stocks. *ICLARM Newsletter* 2(3): 3–4.

Pauly, D. 1986a. A simple method for estimating the food consumption of fish populations from growth data and food conversion experiments. *United States Fishery Bulletin* 4(4): 827–842.

Pauly, D., 1986b. Problems of tropical inshore fisheries: fishery research on tropical soft-bottom communities and the evolution of its conceptual base, pp. 29–37. *In:* E.M. Borgese and N. Ginsburg (eds.), *Ocean Yearbook 6.* University of Chicago Press, Chicago.

Pauly, D. 1988. Fisheries research and the demersal fisheries of Southeast Asia, pp. 329–348. *In:* J.A. Gulland (ed.), *Fish Population Dynamics* (2nd edition). Wiley Interscience, Chichester, UK.

Pauly, D. 1990. On holism, reductionism and working form 9 to 5. *Naga, the ICLARM Quarterly* 13(2): 3–4. [Reprinted as Essay no. 22, p. 159–171 *In:* D. Pauly. 1994. *On the Sex of Fish and the Gender of Scientists: Essays in Fisheries Science.* Chapman & Hall, London.]

Pauly, D. 1994. On Malthusian overfishing, Chapter 13, pp. 112–117. *In: On the Sex of Fish and the Gender of Scientists.* Chapman & Hall, London (UK) [Reprinted from *Naga, the ICLARM Quarterly* 13(1): 3–4.]

Pauly, D. 1995. Anecdotes and the shifting baseline syndrome of fisheries. *Trends in Ecology & Evolution* 10(10): 430.

Pauly, D. 1996. One hundred million tonnes of fish, and fisheries research. *Fisheries Research* 25(1): 25–38.

Pauly, D. 1997. Small-scale fisheries in the tropics: marginality, marginalization and some implication for fisheries management, pp. 40–49. *In:* E.K. Pikitch, D.D. Huppert and M.P. Sissenwine (eds.), *Global Trends: Fisheries Management.* American Fisheries Society Symposium 20, Bethesda, Maryland.

Pauly, D. 1998. Beyond our original horizons: the tropicalization of Beverton and Holt. *Reviews in Fish Biology and Fisheries* 8(3): 307–334.

Pauly, D. 2000. Global change, fisheries, and the integrity of marine ecosystems: the future has already begun, pp. 227–239. *In:* D. Pimentel, L. Westra and R.F. Ross (eds.), *Ecological Integrity: Integrating Environment, Conservation and Health.* Island Press, Washington, DC.

Pauly, D. 2002a. Consilience in oceanographic and fishery research: a concept and some digressions, pp. 41–46. *In:* J. McGlade, P. Cury, K.A. Koranteng and N.J. Hardman-Mountford (eds.), *The Gulf of Guinea Large Marine Ecosystem: Environmental Forcing and Sustainable Development of Marine Resources.* Elsevier Science, Amsterdam.

Pauly, D. 2002b. Review of "The skeptical environmentalist; measuring the real state of the world" by B. Lomborg. *Fish and Fisheries* 3: 3–4.

Pauly, D. 2003. Lifeline: Daniel Pauly. *Nature* 421: 23.

Pauly, D. 2004a. Empty nets. *Alternatives: Canadian Environmental Ideas and Actions* 30(2): 8–13.

Pauly, D. 2004b. Darwin's Fishes: An Encyclopedia of Ichthyology, Ecology and Evolution. Cambridge University Press, Cambridge. xxv+340p.

Pauly, D. 2005. The marine trophic index: a new output of the *Sea Around Us* website. *Sea Around Us Project Newsletter* May/June (29): 1–3.

Pauly, D. 2006a. Unsustainable marine fisheries. *Sustainable Development Law & Policy* 7(1): 10–12, 79.

Pauly, D. 2006b. Babette's Feast in Lima. *Sea Around Us Project Newsletter* November/December (38): 1–2.

Pauly, D. 2007. The *Sea Around Us* Project: Documenting and Communicating Global Fisheries Impacts on Marine Ecosystems. *AMBIO: a Journal of the Human Environment* 34(4): 290–295.

Pauly, D. 2008. Agreeing with Daniel Bromley. *Maritime Studies (MAST)* 6(2): 27–28.

Pauly. D., J. Alder, A. Bakun, S. Heileman, K.H.S. Kock, P. Mace, W. Perrin, K.I. Stergiou, U.R. Sumaila, M. Vierros, K.M.F. Freire, Y. Sadovy, V. Christensen, K. Kaschner, M.L.D. Palomares, P. Tyedmers, C. Wabnitz, R. Watson and B. Worm. 2005. Marine Systems. Chapter 18, pp. 477–511. *In:* R. Hassan, R. Scholes and N. Ash (eds.), *Ecosystems and Human Well-being: Current States and Trends,* Vol. 1, Chapter 20. Millennium Ecosystem Assessment and Island Press, Washington, DC.

Pauly, D., J. Alder, E. Bennett, V. Christensen, P. Tyedmers and R. Watson. 2003. The Future for Fisheries. *Science* 302: 1359–1361.

Pauly, D., J. Alder, E. Bennett, V. Christensen, P. Tyedmers and R. Watson. 2006. World Fisheries: the Next 50 Years, pp. 29–36. *In:* D. Kennedy and the editors of Science Magazine (eds.), *Science Magazine's State of the Planet, 2006–2007.* AAAS and Island Press, Washington, DC. [Updated reprint from "The Future for Fisheries", *Science* 302: 1359–1361.]

Pauly, D., J. Alder, S. Booth, W.W.L. Cheung, V. Christensen, C. Close, U.R. Sumaila, W. Swartz, A. Tavakolie, R. Watson, L. Wood and D. Zeller. 2008. Fisheries in large marine ecosystems: descriptions and diagnoses, pp. 23–40. *In:* K. Sherman and G. Hempel (eds.), *The UNEP Large Marine Ecosystem Report: a Perspective on Changing Conditions in LMEs of the World's Regional Seas.* UNEP Regional Seas Reports and Studies No. 182.

Pauly, D. and V. Christensen. 1993. Stratified models of large marine ecosystems: a general approach and an application to the South China Sea, pp. 148–174. *In:* K. Sherman, L.M. Alexander and

B.D. Gold (eds.), *Large Marine Ecosystems: Stress, Mitigation and Sustainability.* AAAS Press, Washington, DC.

Pauly, D. and V. Christensen. 1995. Primary production required to sustain global fisheries. *Nature* 374: 255–257. [Erratum in *Nature* 376: 279.]

Pauly, D. and V. Christensen (eds.). 1996. *Mass-Balance Models of North-Eastern Pacific Ecosystems.* Fisheries Centre Research Reports 4(1), 131 p.

Pauly, D. and V. Christensen. 2000. Trophic levels of fish, p. 181 *In:* R. Froese and D. Pauly (eds.), *FishBase 2000: Concepts, Design and Data Sources.* ICLARM, Los Baños, Philippines.

Pauly, D. and V. Christensen. 2002. Ecosystem Models. Chapter 10, p. 211–227 *In:* P. Hart and J. Reynolds (eds.), *Handbook of Fish and Fisheries,* Volume 2. Blackwell Publishing, Oxford.

Pauly, D., V. Christensen, J. Dalsgaard, R. Froese and F. Torres Jr. 1998a. Fishing down marine food webs. *Science* 279: 860–863.

Pauly, D., V. Christensen, S. Guénette, T.J. Pitcher, U.R. Sumaila, C.J. Walters, R. Watson and D. Zeller. 2002. Towards sustainability in global fisheries. *Nature* 418: 689–695.

Pauly, D., V. Christensen and C. Walters. 2000. Ecopath, Ecosim and Ecospace as tools for evaluating ecosystem impact of fisheries. *ICES Journal of Marine Science* 57: 697–706.

Pauly, D. and R. Chuenpagdee. 2003. Development of fisheries in the Gulf of Thailand large marine ecosystem: analysis of an unplanned experiment, pp. 337–354. *In:* G. Hempel and K. Sherman (eds.), *Large Marine Ecosystems of the World: Change and Sustainability.* Elsevier Science, Amsterdam.

Pauly, D., R. Froese and V. Christensen. 1998b. How pervasive is "Fishing Down Marine Food Webs"? *Science* 282(5393): 1383a [online publication].

Pauly, D., W. Graham, S. Libralato, L. Morissette and M.L.D. Palomares. 2009. Jellyfish in ecosystems, online databases and ecosystem models. *Hydrobiologia* 616(1): 67–85.

Pauly, D. and J.L. Maclean. 2003. *In a Perfect Ocean: the State of Fisheries and Ecosystems in the North Atlantic Ocean.* Island Press, Washington, DC, 176 p.

Pauly, D. and M.L. Palomares. 2001. Fishing down marine food webs: an update, pp. 47–56. *In:* L. Bendell-Young and P. Gallaugher (eds.), *Waters in Peril.* Kluwer Academic Publishers. Dordrecht.

Pauly, D. and M.L. Palomares. 2005. Fishing down marine food web: It is far more pervasive than we thought. *Bulliten of Maritime Science* 76(2): 197–211.

Pauly, D., M.L. Palomares, R. Froese, P. Sa-a, M. Vakily, D. Preikshot and S. Wallace. 2001a. Fishing down Canadian aquatic food webs. *Canadian Journal of Fisheries and Aquatic Sciences* 58: 51–62.

Pauly, D. and T.J. Pitcher. 2000. Assessment and mitigation of fisheries impacts on marine ecosystems: a multidisciplinary approach for basin-scale inferences, applied to the North Atlantic, pp. 1–12. *In:* D. Pauly and T.J. Pitcher (eds.), *Methods for Evaluating the Impacts of Fisheries on North Atlantic Ecosystems.* Fisheries Centre Research Reports 8(2).

Pauly, D., M.L. Soriano-Bartz and M.L.D. Palomares. 1993. Improved construction, parameterization and interpretation of steady-state ecosystem models, pp. 1–13. *In:* V. Christensen and D. Pauly (eds.), *Trophic Models of Aquatic Ecosystems.* ICLARM Conference Proceedings 26.

Pauly, D. and K. Stergiou. 2005. Equivalence of results from two citation analyses: Thomson ISI's Citation Index and Google's Scholar Service. *Ethics in Science and Environmental Politics* 2005: 33–35.

Pauly, D. and K.I. Stergiou. 2008. Re-interpretation of "influence weight" as a citation-based Index of New Knowledge (INK). *Ethics in Science and Environmental Politics* 8: 1–4.

Pauly, D., A. Trites, E. Capuli and V. Christensen. 1998c. Diet composition and trophic levels of marine mammals. *ICES Journal of Marine Science* 55: 467–481. [Erratum in *ICES Journal of Marine Science* 55: 1153, 1998.]

Pauly, D., P. Tyedmers, R. Froese and L.Y. Liu. 2001b. Fishing down and farming up the food web. *Conservation Biology in Practice* 2(4): 25.

Pauly, D. and R. Watson. 2005. Background and interpretation of the "Marine Trophic Index" as a measure of biodiversity. *Philosophical Transactions of the Royal Society—Biological Sciences* 360: 415–423.

Pearson, H. 2001 China caught out as model shows net fall in fish. *Nature* 467: 477.

Perry, R.I., C.J. Walters and J.A. Boutillier. 1999. A framework for providing scientific advice for the management of new and developing invertebrate fisheries. *Reviews in Fish Biology and Fisheries* 9(2): 125–150.

Peterson, G.D., G.S. Cumming and S.R. Carpenter. 2003. Scenario planning: a tool for conservation in an uncertain world. *Conservation Biology* 17(2): 358–366.

Pew Oceans Commission. 2003. *America's Living Oceans: Charting a Course for Sea Change*. Pew Oceans Commission, Arlington, VA.

Pike, S.T. and A. Spilhaus. 1962. *Marine Resources: a Report to the Committee on Natural Resources of the National Academy of Sciences—National Research Council*. Publication 1000-E, Washington, DC, 8 p.

Pikitch, E.K., C. Santora, E.A. Babcock, A. Bakun, R. Bonfil, D.O. Conover, P. Dayton, P. Doukakis, D. Fluharty, B. Heneman, E.D. Houde, J. Link, P.A. Livingston, M. Mangel, M.K. McAllister, J. Pope and K.J. Sainsbury. 2004. Ecosystem-based fishery management. *Science* 305: 346–347.

Pimm, S.L. 1982. *Food Webs*. Chapman and Hall, London, 219 p.

Pimm, S.L. 2001. *The World According to Pimm: A Scientist Audits the Earth*. McGraw-Hill, New York. 285 p.

Pinnegar, J.K., S. Jennings, C.M. O'Brien and N.V.C. Polunin. 2002. Long-term changes in the trophic level of the Celtic Sea fish community and fish market price distribution. *Journal of Applied Ecology* 39(3): 377–390.

Pinnegar, J.K., N.V.C. Polunin and F. Badalamenti. 2003. Long-term changes in the trophic level of western Mediterranean fishery and aquaculture landings. *Canadian Journal of Fisheries and Aquatic Sciences* 60(2): 222–235.

Pinnegar, J.K., N.V.C. Polunin, P. Francour, F. Badalamenti, R. Chemello, M. L. Harmelin-Vivien, B. Hereu, M. Milazzo, M. Zabala, G. D'Anna and C. Pipitone. 2000. Trophic cascades in benthic marine ecosystems: lessons for fisheries and protected-area management. *Environmental Conservation* 27(2): 179–200.

Pitcher, T.J. 1998. A cover story: fisheries may drive stocks to extinction. *Reviews in Fish Biology and Fisheries* 8(3): 367–370.

Pitcher, T.J. 2001. Fisheries managed to rebuild ecosystems? Reconstructing the past to salvage the future. *Ecological Applications* 11(2): 601–617.

Pitcher, T.J., D. Kalikoski and G. Pramod (eds.). 2006. *Evaluations of Compliance with the UN Code of Conduct for Responsible Fisheries. Fisheries Centre Research Reports* 14(2), 1192 p.

Pitcher, T.J., D. Kalikoski, G. Pramod and K. Short. 2009a. Not honouring the Code. *Nature* 457: 658–659.

Pitcher, T.J., D. Kalikoski, G. Pramod and K. Short. 2009b. *Safe Conduct? Twelve Years Fishing under the UN Code*. WWF, Gland, Switzerland, 63 p.

Pitcher, T.J., R. Watson, R. Forrest, H. Valtysson and S. Guénette. 2002. Estimating illegal and unreported catches from marine ecosystems: a basis for change. *Fish and Fisheries* 3(4): 317–339.

Polovina, J.J. 1984a. Model of a coral reef ecosystem I. The ECOPATH model and its application to French Frigate Shoals. *Coral Reefs* 3(1): 1–11.

Polovina, J.J. 1984b. An overview of the ECOPATH model. *Fishbyte* (ICLARM) 2(2): 5–7.

Polovina, J.J. 1993. The first Ecopath, pp. vii-viii. *In:* V. Christensen and D. Pauly (eds.), *Trophic Models of Aquatic Ecosystems.* ICLARM Conference Proceedings 26.

Polovina, J.J. and M.D. Ow. 1983. *ECOPATH: A User's Manual and Program Listings.* NOAA National Marine Fisheries Service Southwest Fisheries Center Honolulu Laboratory, Honolulu, HI, 46 p.

Polovina, J.J. and M.D. Ow. 1985. An approach to estimating an ecosystem box model. *United States Fishery Bulletin* 83(3): 457–460.

Poon, A. 2000. Retreat at Dunsmuir Lodge: a successful and productive methodology workshop. *Sea Around Us Newsletter*, May–June 2000, p. 1.

Pope, J. 1989. Fisheries research and management for the North-Sea—the next 100 years. *Dana* 8: 33–43.

Post, W.M., F. Chavez, P.J. Mulholland, J. Pastor, T.H. Peng, K. Prentice and T. Webb. 1992. Climatic feedbacks in the global carbon cycle, pp. 392–412. *In:* D.A. Dunnette and R.J. O'Brien (eds.), *The Science of Global Change: the Impact of Human Activities on the Environment.* American Chemical Society Symposium Series, Washington, Vol. 483.

Power, M.E. 1992. Top-down and bottom-up forces in food webs—do plants have primacy? *Ecology* 73(3): 733–746.

Pramod, G., T.J. Pitcher, J. Pearce and D. Agnew. 2008. *Sources of Information Supporting Estimates of Unreported Fishery Catches (IUU) for 59 Countries and the High Seas Fisheries.* Fisheries Centre Research Report 16(4), 242 p.

Pullin, R.S.V., H. Rosenthal and J.L. Maclean (eds.). 1993. *Environment and Aquaculture in Developing Countries.* ICLARM, Manila, Philippines, ICLARM Conference Proceedings 31, 359 p.

Punt, A.E. 2000. Extinction of marine renewable resources: a demographic analysis. *Population Ecology* 42(1): 19–27.

Rabalais, N.N., R.E. Turner, D. Justić, Q. Dortch, W.J. Wiseman and B.K.S. Gupta. 1996. Nutrient changes in the Mississippi River and system responses on the adjacent continental shelf. *Estuaries and Coasts* 19(2B): 386–407.

Radford, T. 1995. New calculations by scientists show there can be no winners in fight for dwindling catches. *The Guardian* (UK), March 16, 1995.

Rawski, T.G. and W. Xiao. 2001. Roundtable on Chinese economic statistics introduction. *China Economic Review* 12(4): 298–302.

Rees, W. and M. Wackernagel. 1994. Ecological footprints and appropriated carrying capacity: measuring the natural capacity requirements of the human economy, pp. 362–390. *In:* A. Jansson, M. Hammer, C. Folke and R. Costanza (eds.), *Investing in Natural Capital.* Island Press, Washington, DC.

Rees, W. E. and M. Wackernagel. 1996. *Our ecological footprint: reducing human impact on the Earth.* New Society Publishers, Gabriola Island, BC, 176 p.

Reitz, E.J. 2004. "Fishing down the food web": A case study from St. Augustine, Florida, USA. *American Antiquity* 69(1): 63–83.

Rice, J. and H. Gislason. 1996. Patterns of change in the size spectra of numbers and diversity of the North Sea fish assemblage, as reflected in surveys and models. *ICES Journal of Marine Science* 53(6): 1214–1225.

Ricker, W.B. 1969. Food from the Sea pp. 87–108. *In:* P. Cloud (chairman) Resources and man, the report of the committee and man to the U.S. National Academy of Science. W.E. Freeman, San Francisco.

Rigler, F.H. 1975. The concept of energy flow and nutrient flow between trophic levels, pp. 15–26. *In:* W.H. van Dobben and R.H. Lowe-McConnel (eds.), *Unifying Concepts in Ecology.* Dr. W. Junk B.V. Publishers, The Hague.

Ritter, M. 1995. Global Fishing Taking Up Much of Ocean's Algae, Study Says. Associated Press, March 16, 1995.

Roach, J. 2005. Penguins marching slowly toward recovery in Argentina, experts say. *National Geographic News* (http://news.nationalgeographic.com).

Robb, A.P. and J.R.G. Hislop. 1980. The food of 5 gadoid species during the pelagic O-group phase in the Northern North-Sea. *Journal of Fish Biology* 16(2): 199–217.

Roberts, C.M. 2007. *The Unnatural History of the Sea: the Past and Future of Humanity and Fishing.* Island Press, Washington, DC, 456 p.

Roberts, C.M., J.A. Bohnsack, F. Gell, J.P. Hawkins and R. Goodridge. 2001. Effects of marine reserves on adjacent fisheries. *Science* 294: 1920–1923.

Roberts, C.M., C.J. McClean, J.E.N. Veron, J.P. Hawkins, G.R. Allen, D.E. McAllister, C.G. Mittermeier, F.W. Schueler, M. Spalding, F. Wells and C. Vynne. 2002. Marine biodiversity hotspots and conservation priorities for tropical reefs. *Science* 295: 1280–1284.

Roberts, C.M. and N.V.C. Polunin. 1993. Marine reserves: Simple solutions to managing complex fisheries? *AMBIO: a Journal of the Human Environment* 22(6): 363–368.

Roger, C. 1994. The plankton of the tropical western Indian Ocean as a biomass indirectly supporting surface tunas (yellowfin, *Thunnus albacares* and skipjack, *Katsuwonus pelamis*). *Environmental Biology of Fishes* 39: 161–172.

Rose, A. 2008. *Who killed the Grand Banks: the untold story behind the decimation of one of the world's greatest natural resources.* John Wiley & Sons, Mississauga, 320 p.

Rosenberg, A.A., M.J. Fogarty, M.P. Sissenwine, J.R. Beddington and J.G. Shepherd 1993. Achieving sustainable use of renewable resources. *Science* 262: 828–829.

Russ, G.R. 1991. Coral reef fisheries, pp. 601–635. *In:* P.F. Sale (ed.), *The Ecology of Fishes on Coral Reefs.* Academic Press, San Diego.

Russ, G.R. 2002. Yet another review of marine reserves as reef fisheries management tools, pp. 421–443. *In:* P.F. Sale (ed.), *Coral reef fishes: Dynamics and Diversity in a Complex Ecosystem.* Academic Press, San Diego.

Ryther, J. 1969. Photosynthesis and fish production in the sea. *Science* 166: 72–76.

Sadovy, Y.J. 2001. The threat of fishing to highly fecund fishes. *Journal of Fish Biology* 59: 90–108.

Sadovy, Y.J. and A.C.J. Vincent. 2002. Ecological issues and the trades in live reef fishes, pp. 391–420. *In:* P.F. Sale (ed.), *Coral Reef Fishes: Dynamics and Diversity in a Complex Ecosystem.* Academic Press, San Diego.

Sainsbury, K.J., A.E. Punt and A.D.M. Smith. 2000. Design of operational management strategies for achieving fishery ecosystem objectives. *ICES Journal of Marine Science* 57(3): 731–741.

Sala, E., O. Aburto-Oropeza, M. Reza, G. Paredes and L.G. Lopez-Lemus. 2004. Fishing down coastal food webs in the Gulf of California. *Fisheries* 29(3): 19–25.

Sanchez, F. and I. Olaso. 2004. Effects of fisheries on the Cantabrian Sea shelf ecosystem. *Ecological Modelling* 172(2–4):151–174.

Sarmiento, J.L., R. Slater, R. Barber, L. Bopp, S.C. Doney, A.C. Hirst, J. Kleypas, R. Matear, U. Mikolajewicz, P. Monfray, V. Soldatov, S.A. Spall and R. Stouffer. 2004. Response of ocean ecosystems to climate warming. *Global Biogeochemical Cycles* 18: GB3003; doi:10.1029/2003GB002134.

Schaefer, M.B. 1954. Some aspects of the dynamics of populations important to the management of the commercial marine fisheries. Bulletin. *Inter-American Tropical Tuna Commission* 1(2): 27–56.

Schaefer, M.B. 1965. The potential harvest of the sea. *Transactions of the American Fisheries Society* 94: 123–128.

Schaefer, M.B. and D.L. Alverson. 1968. World fish potentials, pp. 81–85 *In: The Future of the Fishing Industry of the US.* University of Washington Fisheries Publications (New Series), Vol. 4, Seattle, WA.

Scheffer, M., S. Carpenter, J.A. Foley, C. Folke and B. Walker. 2001. Catastrophic shifts in ecosystems. *Nature* 413: 591–596.

Shaffer, G., S. Malskær Olsen and J.O. Pepke-Pedersen. 2009. Long-term ocean oxygen depletion in response to carbon dioxide emissions from fossil fuels. *Nature Geoscience* 2: 105–109.

Sheldon, R.W., W.H. Sutcliffe Jr. and M.A. Paranjape. 1977. The structure of the pelagic food chain and the relationship between plankton and fish production. *Journal of the Fisheries Research Board of Canada* 34: 2344–2353.

Sherman, K. and G. Hempel (eds.). 2008. *The UNEP Large Marine Ecosystem Report: a Perspective on Changing Conditions in LMEs of the World's Regional Seas. UNEP Regional Seas Reports and Studies No. 182,* 852 p.

Sibert, J., J. Hampton, P. Kleiber and M. Maunder. 2006. Biomass, size, and trophic status of top predators in the Pacific Ocean. *Science* 314: 1773–1776.

Silvestre, G. and D. Pauly. 1997a. Management of Tropical Coastal Fisheries in Asia: an overview of key challenges and opportunities, pp. 8–25. *In:* G. Silvestre and D. Pauly (eds.), *Status and Management of Tropical Coastal Fisheries in Asia.* ICLARM Conference Proceedings 53.

Silvestre, G. and D. Pauly (eds.). 1997b. *Status and Management of Tropical Coastal Fisheries in Asia.* ICLARM Conference Proceedings 53.

Slobodkin, L.B. 1961. *Growth and Regulation of Animal Populations.* Holt, Rinehart and Winston, New York, 184 p.

Sokal, R.R. and F.J. Rohlf. 1995. *Biometry: The Principles and Practice of Statistics in Biological Research.* W.H. Freeman Company, New York, 877 p.

Spalding, M.D. and A.M. Grenfell. 1997. New estimates of global and regional coral reef areas. *Coral Reefs* 16: 225–230.

Spalding, M.D., C. Ravilious and E.P. Green. 2001. *World Atlas of Coral Reefs.* University of California Press, Berkeley, 424 p.

Steele, J.H. 1998. Regime shifts in marine ecosystems. *Ecological Applications* 8 (Suppl. 1): S33-S36.

Steeman-Nielsen, E. 1960. Productivity of the ocean. *Annual Review of Plant Physiology* 11: 341–362.

Steneck, R.S., J. Vavrinec and A. V. Leland. 2004. Accelerating trophic-level dysfunction in kelp forest ecosystems of the Western North Atlantic. *Ecosystems* 7(4): 323–332.

Stephens, D.W. and J.R. Krebs. 1986. *Foraging Theory.* Princeton University Press, Princeton, New Jersey, xiv+247 p.

Stergiou, K.I. 2002. Overfishing, tropicalization of fish stocks, uncertainty and ecosystem management: resharpening Ockham's razor. *Fisheries Research* 55(1–3): 1–9.

Stergiou, K.I. 2005. Fisheries impact on trophic levels: long-term trends in Hellenic waters, pp. 326–329. *In:* E. Papathanassiou and A. Zenetos (eds.), *State of the Hellenic Marine Environment.* Hellenic Centre for Marine Research, Institute of Oceanography, Athens, Greece.

Stergiou, K.I. and M. Koulouris. 2000. Fishing down the marine food webs in the Hellenic seas, pp. 73–78. *In:* F. Briand (ed.), *Fishing Down the Mediterranean Food Webs?* CIESM Workshop Series 12, Kerkyra, Greece.

Stergiou, K.I., A.C. Tsikliras and D. Pauly. 2009. Farming up the Mediterranean food webs. *Conservation Biology* 23(1): 230–232.

Stevens, W. 1998. Man Moves Down the Marine Food Chain, Creating Havoc. *The New York Times,* February 10, 1998.

Strathmann, R.R. 1967. Estimating the organic carbon content of phytoplankton from cell volume and plasma volume. *Limnology and Oceanography* 12: 411–418.

Sumaila, U.R. 1997. Cooperative and non-cooperative exploitation of the Arcto-Norwegian cod stock in the Barents Sea. *Environmental and Resource Economics* 10: 147–165.

Sumaila, U.R. 1999. Pricing down marine food webs, p. 87. *In:* D. Pauly, V. Christensen and L. Coelho (eds.), *Proceedings of the Expo '98 Conference on Ocean Food Webs and Economic Productivity, Lisbon, Portugal, 1–3 July 1998.* ACP-EU Fisheries Research Reports (5), Brussels.

Sumaila, U.R. and M. Bawumia. 2000. Ecosystem justice and the marketplace, pp. 140–153. *In:* H. Coward, R. Ommer and T.J. Pitcher (eds.), *Fish Ethics: Justice in the Canadian Fisheries*, Institute of Social and Economic Research (ISER), Memorial University, St. John's, Newfoundland, Canada.

Sumaila, U.R., A. Khan, R. Watson, G. Munro, D. Zeller, N. Baron and D. Pauly. 2007. The World Trade Organization and Global Fisheries Sustainability. *Fisheries Research* 88: 1–4.

Sumaila, U.R., T.J. Pitcher, N. Haggan and R. Jones. 2001. Evaluating the benefits from restored ecosystems: a back to the future approach, pp. 1–7, Chapter 18. *In:* R.S. Johnston and A.L. Shriver (eds.), *Proceedings of the 10th International Conference of the International Institute of Fisheries Economics & Trade*, Corvallis, Oregon, USA.

Sverdrup, H.U., M.W. Johnson and R.H. Fleming. 1946. *The Oceans, Their Physics, Chemistry, and General Biology.* Prentice-Hall, New York, 1087 p.

Tang, Q. 1989. Changes in the biomass of the Yellow Sea ecosystem, pp. 7–35. *In:* K. Sherman and L.M. Alexander (eds.), *Biomass Yields and Geography of Large Marine Ecosystems.* AAAS Selected Symposium 111. Westview Press, Boulder.

Tatara, K. 1991. Utilization of the biological production in eutrophicated sea areas by commercial fisheries and the environmental quality standard for fishing ground. *Marine Pollution Bulletin* 23: 315–319.

Thomson, H. 1951. Latent fisheries resources and means for their development, pp. 28–38 *In:* Proceedings of the United Nation conference on conservation and utilization of resources, Vol. 7.

Thorpe, A. and E. Bennett. 2001. Globalisation and the sustainability of world fisheries: a view from Latin America. *Marine Resource Economics* 16: 143–164.

Thurow, L. 2007. A Chinese Century? Maybe It's the Next One. *The New York Times,* August 19, 2007.

Tiews, K., P. Sucondharmarn and A. Isarankura. 1967. On changes in the abundance of demersal fish stocks in the Gulf of Thailand from 1963/1964 to 1966 as a consequence of the trawl fisheries development. *Marine Research Laboratory Bangkok*, Contribution No 8, 39 p.

Tudela, S., M. Coll and I. Palomera. 2005. Developing an operational reference framework for fisheries management based on a two-dimensional index on ecosystem impact. *ICES Journal of Marine Science* 62: 585–591.

Tufte, E.R. 1983. *The Visual Display of Quantitative Information.* Graphic Press, Cheshire, CT. 199 p.

Tupper, M.H., K. Wickstrom, R. Hilborn, C.M. Roberts, J.A. Bohnsack, F. Gell, J.P. Hawkins and R. Goodridge. 2002. Marine reserves and fisheries management. *Science* 295: 1233–1235.

Turner, S.J., S.F. Thrush, J.E. Hewitt, V.J. Cummings and G. Funnell. 1999. Fishing impacts and the degradation or loss of habitat structure. *Fisheries Management and Ecology* 6(5): 401–420.

Tyedmers, P. 2004. Fisheries and energy use, pp. 683–693. *In:* C. Cleveland (ed.), *Encyclopedia of Energy.* Academic Press/Elsevier, San Diego, CA, Vol. 2.

Ulanowicz, R.E. 1995. Ecosystem Trophic Foundations: Lindeman Exonerata, pp. 549–560. *In:* B.C. Patten and S.E. Jørgensen (eds.), *Complex Ecology: The Part-Whole Relation in Ecosystems.* Prentice-Hall, Englewood Cliffs, NJ.

Ulanowicz, R.E. and C.J. Puccia. 1990. Mixed trophic impacts in ecosystems. *Coenoses* 5: 7–16.

UN. 1982. *The United Nations Convention on the Law of the Sea.* The United Nations, New York.

UNCLOS. 1994. United Nations Convention on the Law of the Sea, http://www.un.org/Depts/los/index.htm.

UNEP. 2002. *Global Environment Outlook 3: Past, Present and Future Perspectives.* United Nations Environment Programme and Earthscan Publications, London, 480 p.

Valtysson, H. 2001. The Sea Around Icelanders: catch history and discards in Icelandic waters, pp. 52–87. *In:* D. Zeller, R. Watson and D. Pauly (eds.), *Fisheries Impacts on North Atlantic Ecosystems: Catch, Effort and National/Regional Data Sets.* Fisheries Centre Research Reports 9(3).

Valtysson, H. and D. Pauly. 2003. Fishing down the food web: an Icelandic case study, pp. 12–24. *In:* E.V. Guðmundsson (ed.), *Proceedings of a Conference held in Akureyri, Iceland, on April 6–7th 2000 Competitiveness within the global fisheries.* University of Akureyri, Akureyri.

van Dolah, F.M., D. Roelke and R.M. Greene. 2001. Health and ecological impacts of harmful algal blooms: risk assessment needs. *Human and Ecological Risk Assessment* 7(5): 1329–1345.

Venema, S., J. Möller-Christensen and D. Pauly (eds.). 1988. *Contributions to tropical fisheries biology: papers by the participants of FAO/DANIDA follow-up training courses.* FAO Fisheries Report No. 389. 519 p.

Veridian Information Solutions, 2000. *Global Maritime Boundaries Database CD.* Veridian, Fairfax, Virginia.

Vierros, M. and D. Pauly. 2004. Assessing biodiversity loss in the oceans: a collaborative effort between the Convention on Biological Diversity and the *Sea Around Us* Project. *Sea Around Us Project Newsletter* May/June (23): 1–4.

Vitousek, P.M., P.R. Ehrlich and A.H. Ehrlich. 1986. Human appropriation of the products of photosynthesis. *Bioscience* 36: 368–373.

Vivekanandan, E., M. Srinath and S. Kuriakose. 2005. Fishing the marine food web along the Indian coast. *Fisheries Research* 72: 241–252.

von Baeyer, H.C. 1993. *The Fermi Solution: Essays on Science.* Random House, New York, 176 p.

von Brandt, A. 1984. *Fish Catching Methods of the World.* Fishing News Books, Oxford. 432 p.

Wackernagel, M. and W. Rees. 1996. *Our Ecological Footprint: Reducing Human Impact on the Earth.* New Society Publishers, Gabriola Island, BC.

Wallace, S. 1999. *Fisheries impacts on marine ecosystems and biological diversity: the role for marine protected areas in British Columbia.* PhD dissertation, University of British Columbia, 198 p.

Walters, C., V. Christensen and D. Pauly. 1997. Structuring dynamic models of exploited ecosystems from trophic mass-balance assessments. *Reviews in Fish Biology and Fisheries* 7(2): 139–172.

Walters, C. and J.F. Kitchell. 2001. Cultivation/depensation effects on juvenile survival and recruitment: implications for the theory of fishing. *Canadian Journal of Fisheries and Aquatic Sciences* 58: 39–50.

Walters, C. and J.J. Maguire. 1996. Lessons for stock assessment from the northern cod collapse. *Reviews in Fish Biology and Fisheries* 6(2): 125–137.

Walters, C., D. Pauly and V. Christensen. 1999. Ecospace: Prediction of mesoscale spatial patterns in trophic relationships of exploited ecosystems, with emphasis on the impacts of marine protected areas. *Ecosystems* 2(6): 539–554.

Walters, C., D. Pauly, V. Christensen and J.F. Kitchell. 2000. Representing density dependent consequences of life history strategies in aquatic ecosystems: EcoSim II. *Ecosystems* 3(1): 70–83.

Ward, M. 2004. *Quantifying the World: UN Ideas and Statistics.* Indiana University Press, Bloomington, 329 p.

Watling, L. and E.A. Norse. 1998. Disturbance of the seabed by mobile fishing gear: A comparison to forest clearcutting. *Conservation Biology* 12(6): 1180–1197.

Watson, R. 2001. Spatial allocation of fisheries landings from FAO Areas 61 and 71, pp. 28–49. *In:* R. Watson, L. Pang and D. Pauly (eds.), *The Marine Fisheries of China: Development and Reported Catches.* Fisheries Centre Research Report 9(2).

Watson, R., A. Gelchu and D. Pauly. 2001b. Mapping fisheries landings with emphasis on the North Atlantic, pp. 1–11. *In:* D. Zeller, R. Watson and D. Pauly (eds.), *Fisheries Impact on North Atlantic Marine Ecosystems: Catch, Effort and National and Regional Data Sets.* Fisheries Centre Research Reports 9(3).

Watson, R., A. Kitchingman, A. Gelchu and D. Pauly. 2004. Mapping global fisheries: sharpening our focus. *Fish and Fisheries* 5: 168–177.

Watson, R., L. Pang and D. Pauly. 2001a. *The Marine Fisheries of China: Development and Reported Catches.* Fisheries Centre Research Report 9(2), 50 p.

Watson, R. and D. Pauly. 2001. Systematic distortions in world fisheries catch trends. *Nature* 414: 534–536.

WCED. 1987. *Our Common Future: Report of the World Commission on Environment and Development.* Oxford University Press, Oxford.

Weber, P. 1995. Facing limits to oceanic fisheries: Part II: The social consequences. *Natural Resources Forum* 19(1): 39–46.

Weiss, K. 2007. The rise of slime: Red tides, jelly-fish plagues, explosions of primitive organisms. *New Internationalist* (397): 6–8 [Reprinted from the *Los Angeles Times.*]

Willemse, N. and D. Pauly. 2004a. Ecosystem overfishing: a Namibian case study, pp. 253–260. *In:* P. Chavance, M. Ba, D. Gascuel, M. Vakily et D. Pauly (eds.), *Pêcheries maritimes, écosystèmes et sociétés en Afrique de l'Ouest: un demi-siècle de changement.* Actes du symposium international, Dakar—Sénégal, 24–28 juin 2002. Office des publications officielles des communautés Européennes, XXXVI, collection des rapports de recherche halieutique ACP-UE 15.

Willemse, N.E. and D. Pauly. 2004b. Reconstruction and interpretation of marine fisheries catches from Namibian waters, 1950 to 2000, pp. 99–112. *In:* U.R. Sumaila, D. Boyer, M.D. Skogen and S.I. Steinshamm (eds.), *Namibia's Fisheries: Ecological, Economic and Social Aspects.* Eburon Academic Publishers, Amsterdam.

Williams, N. 1998. Overfishing disrupts entire ecosystems. *Science* 279: 809 p.

Wilson, E.O. 1998. *Consilience: the Unity of Knowledge.* Knopf, New York, 367 p.

Winberg, G.G. (ed.). 1971. *Method for the Estimation of Production of Aquatic Organisms.* Academic Press, London, 174 p.

Wing, E.S. 2001. The sustainability of resources used by Native Americans on four Caribbean islands. *International Journal of Osteoarchaeology* 11: 112–126.

Wing, S.R. and E.S. Wing. 2001. Prehistoric fisheries in the Caribbean. *Coral Reefs* 20(1): 1–8.

Wood, L., L. Fish, J. Laughren and D. Pauly. 2008. Assessing progress towards global marine protection targets: shortfalls in information and action. *Oryx* 42(3): 340–351.

World Bank. 2009. *The Sunken Billions: the Economic Justification for Fisheries Reform.* The International Bank for Reconstruction and Development. Washington, DC, 100 p.

Worm, B., E.B. Barbier, N. Beaumont, J.E. Duffy, C. Folke, B.S. Halpern, J.B.C. Jackson, H.K. Lotze, F. Micheli, S.R. Palumbi, E. Sala, K.A. Selkoe, J.J. Stachowicz and R. Watson. 2006. Impacts of biodiversity loss on ocean ecosystem services. *Science* 314: 787–760

Yan, H.Y. 2001. China, a country good at statistical falsification. *Liberty Times*, Taipei, Taiwan, December 8, 2001. [In Chinese.]

Yellen, J.E., A.S. Brooks, E. Cornelissen, M.J. Mehlman and K. Stewart. 1995. A middle stone-age worked bone industry from Katanda, Upper Semliki Valley, Zaire. *Science* 268: 553–556.

Yoon, C.K. 2003. Scientists at work: Daniel Pauly—iconoclast looks for fish and finds disaster. *The New York Times,* January 2, 2003.

Zaitsev, Y.P. 1992. Recent changes in the trophic structure of the Black Sea. *Fisheries Oceanography* 1(2): 180–189.

Zeller, D.C. 1997. Home range and activity patterns of the coral trout *Plectropomus leopardus* (Serranidae). *Marine Ecology Progress Series* 154: 65–77.

Zeller, D. and D. Pauly. 2005. Good news, bad news: Global fisheries discards are declining, but so are total catches. *Fish and Fisheries* 6: 156–159.

Zeller, D. and D. Pauly (eds.). 2007. Reconstruction of marine fisheries catches for key countries and regions (1950–2005). *Fisheries Centre Research Reports* 15(2): 163.

Zeller, D., S. Booth, G. Davis and D. Pauly. 2007. Re-estimation of small-scale for U.S. flag-associated islands in the western Pacific: the last 50 years. *Fisheries Bulletin* 105: 266–277.

Index

Figures/photos/illustrations are indicated by an " f " and tables by a " t ."